THE MACHINE IN THE TEXT

The Machine in the Text

Science and Literature in the Age of Shakespeare and Galileo

HOWARD MARCHITELLO

OXFORD

UNIVERSITY PRESS

OXFORD
UNIVERSITY PRESS

Great Clarendon Street, Oxford OX2 6DP

Oxford University Press is a department of the University of Oxford.
It furthers the University's objective of excellence in research, scholarship,
and education by publishing worldwide in

Oxford New York

Auckland Cape Town Dar es Salaam Hong Kong Karachi
Kuala Lumpur Madrid Melbourne Mexico City Nairobi
New Delhi Shanghai Taipei Toronto

With offices in

Argentina Austria Brazil Chile Czech Republic France Greece
Guatemala Hungary Italy Japan Poland Portugal Singapore
South Korea Switzerland Thailand Turkey Ukraine Vietnam

Oxford is a registered trade mark of Oxford University Press
in the UK and in certain other countries

Published in the United States
by Oxford University Press Inc., New York

© Howard Marchitello 2011

British Library Cataloguing in Publication Data

Data available

Library of Congress Cataloging in Publication Data

Data available

Typeset by SPI Publisher Services, Pondicherry, India
Printed in Great Britain
on acid-free paper by
MPG Books Group, Bodmin and King's Lynn

ISBN 978-0-19-960805-8

3 5 7 9 10 8 6 4 2

To Max Vallone Marchitello
and
Rosalie Vallone Marchitello
—*Perché l'arancia . . . ?*

Acknowledgments

Unlike the book they precede, acknowledgments write themselves. And, both needless to say and emphatically, it is because of the many kindnesses and assistance of those individuals and institutions thanked here that this particular book was written at all.

I have had the very good fortune of enjoying the support of not one, but *two* institutions through the years of research and writing this project. The English Department at Texas A&M University was an unfailingly stimulating and congenial setting for work and for play for the better part of two decades. Work on this project benefited from the generous support of J. Lawrence Mitchell (my long-time head of department) and the Dean of Liberal Arts, Charles Johnson—support that took the form of a number of teaching reductions and (in 2004) a Faculty Development leave, which together provided sustained time for concentrated work. I am also deeply—and enduringly—indebted to the remarkable group of scholars and friends affiliated with the Glasscock Center for Humanities Research, in particular James Rosenheim, Margaret J. M. Ezell, Larry J. Reynolds, and Jeffrey N. Cox (now of the University of Colorado, Boulder). Texas A&M also provided material and, just as important, spiritual support in the form of its remarkable study abroad program in association with ITALART and the Santa Chiara Study Center in Castiglion Fiorentino, Italy—a cultural association, under the visionary leadership of Paolo Barucchieri, that has been a welcoming and enriching home to me and my family on many occasions over these last fifteen years. I note with special pleasure the energetic and fearless members of the Spring 2007 Galileo Seminar.

My colleagues at Rutgers University Camden have provided a happy new home. I would like here to thank in particular our Dean, Margaret Marsh, who even a newcomer such as I can see has succeeded in establishing the College on a whole new level of success and promise; I thank her for her support—both moral and material. The College has also provided me with funding for a number of research trips to London and the British Library that have enabled me to complete the final research for this book. My colleagues in the English Department at Camden have been most welcoming: they have endured what must seem an endless stream of procedural and institutional questions; they have lent their ears—and their insights—to presentations of parts of this project, especially the material on C. P. Snow and the two cultures; and they have helped me find my way and find my place in a new institution.

The opportunity to present work in progress and to engage in discussions with colleagues is crucial to any research undertaking. In this regard, I have benefitted immeasurably from such opportunities, including, especially, numerous panels and seminars at both the MLA annual conference and the Shakespeare Association of America conferences over the past several years. I am also pleased to extend my thanks to the Rutgers University Program in Early Modern Studies (headed by Henry S. Turner) and the Center for Cultural Analysis, for inviting me to participate in the Spring 2009 symposium, "Literature and Science in the Early Modern Period."

On occasion, conference papers and presentations lead to publications. Portions of some of the chapters in this book were published in preliminary form in the following venues: work on early modern gardens and John Evelyn's garden practices in "Garden frisson," *JMEMS: Journal of Medieval and Renaissance Studies*, 33/1 (2003): 143–77; early modern literature and science in "Science studies and English Renaissance literature," *Literature Compass*, 3/3 (2006): 341–65; and Shakespearean theater and artificial (or prosthetic) experience in "Artifactual knowledge in *Hamlet*" in *Knowing Shakespeare: Senses, Embodiment and Cognition*, edited by Lowell Gallagher and Shankar Raman (Palgrave Macmillan, 2010).

I am also pleased to extend my thanks to Jacqueline Baker, at Oxford University Press, for her strong support of this project at every stage. I am greatly indebted, too, to the extraordinarily thoughtful and demanding anonymous readers for the press; this book has been vastly improved as a consequence of the rigor of their evaluations and suggestions.

This book has been made better through my conversations with a number of friends and scholars of the early modern period over the course of many years; for their assistance and for their scholarship I would like to thank in particular Lowell Gallagher, Carla Mazzio, Shankar Raman, Elizabeth Spiller, and Henry S. Turner. I would like to record here, as well, my general indebtedness: to Stephen Orgel, for his friendship and for the views of the *Campo*; to David E. Johnson for (among many such stretching back nearly twenty-five years) an especially key conversation on a beautiful Buffalo summer day on the feasibility of books; to Gordon McMullan, for generosity both intellectual and personal that (like the black swan in Essex) seemingly knows no bounds.

My thanks are due also to my parents, Howard and Jane Marchitell, for literally life-long support. Thanks also to John and Phyllis Vallone for their ten thousand kindnesses; to Peter Hunt ("Books Sorted-Out"); to Dan Cook, for the many coffees on campus—and on improbably short notice—over these last few years; and to two Texas friends: Larry J. Reynolds and Dennis Berthold. To Larry, who is a model of intellectual integrity and rigor, thanks for his enduring friendship and for the example his work provides; and thanks to my great friend Dennis, from whom, I am proud to say, I continue to learn—about the profession, about the *Risorgimento*, and about that great adventure more commonly known as "life." Thanks, too, and most recently, for Crazy Horse.

One person has been at the center of my life in every way for almost thirty years. To Lynne Vallone—teacher, scholar, triathelete, best reader, best friend—who not only demonstrates but also enacts, as her scholarship (for instance) makes clear, the truth of *multum in parvo*, I can only offer my deep thanks and love in return.

Finally: *i ragazzi*. What can I say about my son and daughter that our countless hours and countless adventures together across three continents haven't already said? They have grown into fine adults—even before my very eyes: This minute they are born; this minute they are grown. As a sign of my happiness at the very fact of them, and as a token of the love I have for them, I dedicate this book to Max and Rose.

Contents

List of Illustrations

1

Introduction

Science Studies and Early Modern Literature and Culture

The Machine in the Text is intended as a contribution to the ongoing efforts to locate both early modern science and early modern literature more precisely within the rich context of early modern culture, without extracting either one or the other and privileging it in a created, but illusory, isolation. At the same time, I am interested in examining with similar rigor the respective objects of study posited by *scientia* and *ars*: the material world we inhabit and the artificial representations of and engagements with it we create in order (at least in part) to understand it. A set of questions that therefore motivates this book asks about nature and art and the relationship between them in the early modern period. If new science studies-inflected criticism of early modern culture allows us to see that art is a knowledge practice and, conversely, that scientific practices can be understood as imaginative, creative, and literary, then what might we learn from a coordinated investigation of the objects these two interrelated discourses identify as their special concerns?

These (and related) questions have been central to recent work in the field of early modern studies dedicated to rethinking conventional understandings of science, particularly in its relationship to literature (and to culture more generally). The first task of this work has been to unseat the structural formulation of a more or less complete and allegedly natural separation of *scientia* from *ars* that has been profoundly influential for much of the history of the West, but that until quite recently has also gone largely unquestioned. Perhaps the most visible and, in some measure, enduring commentary on the subject has also served to give a name, if not a local habitation, to the entire matter, even if the ensuing debates have produced rather more heat than light: the now commonplace designation of the *two cultures*, a heuristic that comes to us from C. P. Snow's book of the same title (a book—and a phenomenon—that I address in the final chapter of this study). But for the moment, it is sufficient to say that Snow's *Two Cultures* is important less as an analysis than as a symptom of the more fundamental separation of *scientia* and *ars* into strictly enforced disciplinary isolation that was Snow's declared subject and that stands as the purpose of the present work to contest.

Citing conceptual developments in recent and contemporary work in the history and philosophy of science that have "transformed the intellectual landscape for those interested in understanding points of consilience as well as contestations between literature and science, particularly in periods before the rise of the modern disciplines," Carla Mazzio has recently highlighted the new and innovative ways that scholars now study "the distinctly narrative, literary, and hypothetical dimensions of experimental science and, conversely, the scientific dimensions of the literary or fictional experiment."[1] As Mazzio notes (and as I will argue below), at the same time—and in some measure as a consequence of these new approaches and practices in early modern studies—not only is the separation of *scientia* and *ars* erased, but so too is the traditional privileging of science. Historians and literary critics have "rightly challenged earlier assumptions that once posited science as autonomous, proto-rationalist, and privileged with relationship to truth, and accordingly, models of interpretation that found literature merely 'reflective' of scientific principles, or vice versa" (p. 4).

Mazzio makes these comments in her introduction to a journal issue dedicated to "Shakespeare and Science," as a means to locate a diverse set of individual essays that address the relationship between Shakespeare and early modern science. But even given the wide range of topics these essays address—anatomy and cartography in *King Lear* and the engendering of norms; botanical Shakespeares; air and instrumentality in *Hamlet*; Shakespeare and physics; and posthuman Shakespeare—all of them have in common this first necessary step of critical erasure. In her essay in the issue "Shakespeare and the making of early modern science: Resituating Prospero's art," Elizabeth Spiller identifies the early modern period as a moment of monumental transition from Aristotelian *scientia* to modern science and identifies, moreover, the critical role played by art in this transition: "Both classic and modern epistemologies of science exclude art," Spiller writes, "but the crucial transition from the first to second was itself largely achieved by art. Art unexpectedly became the mediating term that made it possible for early modern intellectual culture to abandon Aristotelian scholasticism and move toward experimentalism and fact-based knowledge models."[2] Spiller's central concern in her essay is to examine the ways in which art functioned "as a knowledge practice," even if "the Renaissance conception of art as knowledge was ultimately displaced by a modern science of facts" (p. 25). Based upon these newly formulated understandings of the intimate relation between *scientia* and *ars*, science and art (what I will later expand to include nature and art), Spiller's new reading of Prospero's art is a powerful expression of the theoretical argument that "in order to understand how poetry and drama shared in the emergent scientific cultures of early modern England we must recognize that art was not separate from the practices that became science but instrumental to them" (p. 25).

Contemporary criticism such as Spiller's rereading of Prospero's art, together with the rich variety of work sampled in Mazzio's "Shakespeare and Science," represent the leading edge of a new (or perhaps renewed) interest in literature and science in the early modern period. At the same time, this is work heavily indebted to the field of science studies more generally and to a significant degree can be said to require, among other things, a new discussion of disciplinarity if we are to understand the consequence for the study of literature of the reconceptualization of

---

[1] Mazzio (2009: 4).   [2] Spiller (2009: 24).

art as a knowledge practice within emergent cultures of early modern science. The erasure of the *scientia/ars* polarization that underwrites this new work is a methodological innovation and has yielded great new insights into early modern literature and culture, including, of course, science "itself."

The broad goal of the following introductory pages is to offer a brief overview of the history of science as it impacts the study of early modern literature and culture. This history has two general phases. The first, which begins in earnest in the science and literature criticism of the 1930s and lasts well into the 1980s, is dominated by what I will call the influence model in which the issue most pressingly at stake is the labor of demonstrating the influence of science on literary texts in the early modern English tradition. The second phase, the new wave of science and literature criticism, begins to emerge in the 1980s and 1990s and to develop in more or less direct reaction to a certain revolution in science studies, a term that I will use to designate the multidisciplinary study of science as both a socially and an historically embedded set of practices and habits of thought. This revolution in science studies will also therefore be part of the story this chapter will tell about early modernity, science, and literary culture. The most significant consequence of second-wave science and literature criticism is a new understanding of the ways in which both the scientific and the literary are equally (though, of course, differently) engaged in the production of knowledge.

Another concern of this discussion will be to investigate the nature of pre-disciplinary science, by which I mean science that is still embedded within a range of cultural practices. As such, I will seek to avoid thinking about science as an already autonomous feature of early modernity; rather, attention will be focused on the practices of science in the process, as it were, of their separation from the various cultural sites that in this view constitute less points of origin than sites of emergence. In this, I am following the lead of French historian and philosopher of science, Michel Serres. Serres begins the introduction to a recently published anthology of essays, *A History of Scientific Thought*, by noting a disturbing problem: given the fact that we live in a world "dominated by science and technology" and therefore are increasingly likely "[to] question the whys and wherefores of its recent advent and sometimes even its legitimacy," it is a "great paradox that in schools and universities the history of science is not taught in the same way as the usual disciplines: it is only haphazard, depending on the inclination of the individual teacher."[3] Serres continues:

> We generally learn about our history in isolation from that of the sciences. We study philosophy devoid of any scientific reasoning and great literature in splendid detachment from its scientific context; we study various disciplines uprooted from the soil of their history, as if they had happened by chance. In short, our entire learning process is

---

[3] Serres (1995a: 1). Serres works in and across a wide range of disciplines; René Girard, in his introduction to Serres's book *Detachment* writes, "Michel Serres has written a great deal about scientific discovery, and yet to define him as a historian or even as a philosopher of science would not do justice to the breadth of his work" (vii). For Serres's work that does fall into these fields, see especially (1997), (1995b), (1982), and, with Bruno Latour (1995).

inappropriate to the real world in which we live, a world which is a confused mixture of technology and society, of insane or wise traditions and useful or disturbing innovations. [p. 1]

While these comments reflect an understanding of the nature of the discipline—or, more appropriately, the *disciplines*—of science studies, they also posit a new model for the relationship between the sciences and the rest of society. Serres's metaphor of the "confused mixture" constitutes his model for understanding a relationship in which science is no longer conceived as a separate sphere, as a set of practices that exist at a certain isolating remove from society, as if science were somehow not itself a part of society. Instead, for Serres, both science and the study of science are fully embedded within the social context that produces them.

One significant benefit of this deliberate relocating of science *as an object of study* within the social setting that as an evolving repertoire of practices and objectives science has in fact always inhabited, is to neutralize the deleterious effects of what Serres calls the "cultural crisis caused in part by this estrangement" of science from society: "This divorce between two worlds is sometimes expressed as hostility and sometimes as adoration, both of which are excessive" (p. 2). As Serres's introduction—and his work, more generally—makes clear, the "great paradox" of the study of science is a matter of history, of the ways in which science has come to be either the object of hostility or—perhaps even more problematically—of adoration. It is in this spirit that the following discussion attempts neither to condemn nor to celebrate science, but rather to recover its socially and historically situated complexities.[4]

A final word before beginning the discussion proper: what we typically think of as early modern science is a very broad field indeed that encompasses any number of disciplinary practices, customary objects of study, and particular goals and objectives. And while early modern science can mean medicine, alchemy, anatomy, biology, horticulture, physics, mechanics, optics, mineralogy, geometry, mathematics, astronomy, among many other fields, I will use the term to designate an emergent set of mental practices and habits of thought that can be said to inform those various disciplinary articulations of science that we are used to thinking about now as separate fields of study.

---

[4] This sort of work dedicated to these complexities stands in stark contrast to what Serres calls the "spontaneous history of the sciences, as August Comte would have said." This spontaneous history would have numerous liabilities, including a faith in an "unhampered progress of total knowledge through global, homogenous and isotropic time," a deep commitment "to paint the portraits of the geniuses who made the discoveries." And, "above all," Serres continues, this spontaneous history of science

assumes this retrograde movement of the truth which projects the knowledge of today onto the past in such a way that history irresistibly becomes an almost programmed preparation for the knowledge of today. In truth, there is nothing more difficult than imagining a time that was free and fluctuating and not completely determined . . . (1995a: 5).

I

The notion of sites of emergence, as suggested above, works explicitly against the idea of identifiable instants of origin and does so, in part, in order to avoid the liabilities of a certain teleological notion of progress that often attends narratives of the history of science—a liability that John Henry, in his book *The Scientific Revolution and the Origins of Modern Science*, identifies as Whig history. "There is a tendency," Henry writes, "in the history of science to look back with hindsight about what is known to be important later."[5] Teleological understandings of the nature of science are also enabled by a conception of science as an accumulation of knowledge and admonitions about the dangers of such a notion have been regular features of histories of science for many decades. Charles Singer begins his classic study, *A Short History of Science to the Nineteenth Century* (1941), with a preliminary discussion of just this concern; his introduction ("Nature of the scientific process"), as its title suggests, recommends the rejection of the notion of science as a body of knowledge and its replacement with a model of process:

> Science is often conceived as a *body of knowledge*. Reflection, however, will lead to the conclusion that this cannot be its true nature. History has repeatedly shown that a body of scientific knowledge that ceases to develop soon ceases to be science at all. The science of one age has often become the nonsense of the next. Consider, for example, astrology; or, again, the idea that certain numbers are lucky or unlucky. With their history unknown, who would see in these superstitions the remnants of far-reaching scientific doctrines that once attracted clear-thinking minds seeking rational explanations of the working of the world?[6]

But in spite of this recognition of the processive nature of science (as opposed to the merely accumulative "body of knowledge" argument), Singer slides into an uncritical *progressive* mode in which science inexorably moves forward into time and into greater accuracy. Singer's view is of a self-correcting science governed by an always evolving—which is to say, increasingly perfect—rationality, even if perfection is understood always to be receding into the distance. On this view, what are cast as former sciences (astrology, say), function less as the caution against a certain hubris on behalf of our own scientific knowledges that Singer claims they are, than simply as instances of the finally non-scientific practices of the past. In Singer's account, they become evolutionary dead-ends that serve to authorize the evolutionary model

---

[5] John Henry (1997: 2). Henry continues:

To judge the past in terms of the present is to be whiggish. In the early decades of the formation of the discipline it was common for a historian of science to pick out from, say, Galileo's work, or Kepler's, those features which were, or could most easily be made to look like, direct anticipations of currently held science. The resulting history was often a lamentable distortion of the way things were.

[6] Charles Singer (1941: 1). Singer wrote extensively on the history of science, with particular attention on anatomy, biology, technology, and the history of medicine.

itself: nothing, one could say, authorizes our own scientific culture more than the fact that we no longer believe in astrology as a science.[7]

The triumphalist model of science and its history held by Singer—which was, if not perfectly representative, then at least typical, of the study of science well into the 1970s—was also the presiding model of science at the heart of literature and science studies across a wide range of periods and national traditions. The literary critical work that this understanding of science (and of history) served to prompt was vast and varied, but in general terms took more or less the form of influence studies. One particularly apt example within the field of Renaissance literary studies focused on science—an example of what I would call the *first wave* of science and early modern literature studies—can be found in the work of Marjorie Hope Nicolson. Nicolson published a significant number of important essays and books on Renaissance literature and its relation to science in the 1930s and 1940s.[8] Six of her essays were collected and published in 1956 under the title *Science and Imagination* and as Nicolson announces in the first essay of the collection, "The telescope and the imagination," she understands the relation of science to literature of the period *not* to be a matter of mere influence, but something altogether more complicated and important. Discussing innovations and discoveries in astronomical science (which is the particular focus of all the essays collected in this volume), Nicolson wants to stress the *structural* impact of science on the mind or imagination of the literary artist:

> There is a feeling here of change, of awareness of astronomical implication which both disturbs and fascinates the seventeenth-century mind. On the one hand, man is shrinking back from an unknown gulf of immensity, in which he feels himself swallowed up; on the other, he is, like [Giordano] Bruno, "rising on wings sublime" to a spaciousness of thought he had not known before. The poetic and religious imagination of the century was not only influenced, but actually changed, by something latent in the "new astronomy."[9]

It becomes clear immediately, however, that Nicolson's argument about the literature–science relationship is a fundamentally asymmetrical one in which the

---

[7] The very structure of Singer's book carries and delivers its own potent portion of this evolutionary argument: "Chapter 1. Rise of Mental Coherence. The Foundations (600–400 BC): Ionia, Magna Graecia, Athens." "Chapter 2. The Great Adventure. Unitary Systems of Thought (400–300 BC): Athens." "Chapter 3. The Failure of Nerve. Divorce of Science and Philosophy (300 BC–AD 200): Alexandria." These chapters are followed by two additional chapters recounting failures, Chapter 4, subtitled "Science the Handmaid of Practice," on Imperial Rome, and Chapter 5 on the Middle Ages and subtitled "Theology, Queen of Sciences." But, true to the progressive model (which also happens to have been essentially the same model practiced by many historians of literature), Singer's story begins its move toward a happy ending, with the Renaissance serving as the great pivot: "Chapter 5: The Revival of Learning. The Rise of Humanism (1250–1600). The Attempted Return to Antiquity," "Chapter 7. The Insurgent Century. Downfall of Aristotle (1600–1700). New Attempts at Synthesis," and ending (literally, culminating) in "Chapter 8. The Mechanical World. XVIIIth–XIXth Century. Enthronement of Determinism (1700–19th Century)."

[8] Nicolson wrote widely on seventeenth- and early eighteenth-century writers and poets such as John Donne, John Milton, Samuel Pepys, Alexander Pope, and Isaac Newton; see also (1946) and (1965).

[9] Nicolson (1956: 2).

generative power of the new astronomy powerfully influences, and even changes, the literary imagination of the age: "New figures of speech appear, new themes for literature are found, new attitudes toward life are experienced, even a new conception of Deity emerges" (p. 2). In her discussions of Galileo's 1610 *Sidereus Nuncius* ("the most important single publication, it seems to me, of the seventeenth century, so far as its effect upon imagination is concerned" [p. 4]), for example, and in her more elaborate and sustained reading of John Donne, Nicolson's overall argument, as valuable as it has been, is nevertheless a limited one precisely because it is restricted entirely to the ways in which literary texts were impacted by science. One of the presuppositions that sustains this sort of understanding is an unquestioned belief in the already autonomous nature of science that functions with a virtual agency of its own to transform first the literary imagination, and then the literary texts which that imagination produces. Nicolson's is a one-way street: science influences (or changes) literary production while maintaining a defining and isolating distance. There is never any consideration of the ways in which the literary might be described as influencing (or changing) the scientific. An effect of this interpretive model is to relegate the literary to an essentially *reflective* status because science produces literary consequences:

I shall try at the present time to reconstruct the instantaneous effect the discoveries reported there [in *Sidereus Nuncius*] had upon the poets of Galileo's own country, in order that their effect upon poetic imagination in England may be better understood. In the papers that follow, I shall trace the course of that effect in England, the merging of the Galilean ideas with those already native there, and then shall follow the development of the telescope and microscope as they appear in literature, watching new figures of speech, new literary themes, new cosmic epics, most of all the transformation of poetic and religious imagination by ideas which, once grasped, man has never been able to forget. [p. 4]

One of the particular "figures of speech" Nicolson will address in her discussion of the impact on the "literary imagination" of Galileo's book is the "Columbus-motif" (p. 19) deployed by many poets in their praise of Galileo's discoveries. Among these poets were Thomas Seggett (one of Galileo's British pupils in Padua) and Johannes Faber, whose commendatory poem ("Ad Galilaeum Lynceum Florentinum Mathematicorum...") was eventually published in Galileo's 1623 *Il Saggiatore* (*The Assayer*) and in part reads,

> Yield, Vespucci, and let Columbus yield. Each of these
> Holds, it is true, his way through the unknown sea...
> But you, Galileo, alone gave to the human race the sequence of stars,
> New constellations of heaven.[10]

[10] This is Nicolson's translation of Faber, "Ad Galilaeum Lynceum Florentinum Mathematicorum Saeculi Nostri Principem Mirabilium in Caelo per Telescopium Novum Naturae Oculum Inventorem," (1956: 19). Nicolson also cites the long poem *Adone*, by Giambattista Marino; a sonnet by Piero de' Bardi; as well as Andrea Salvadore's poem "Per le Stelle Medicee," and "Adulatio Pernicisa," written by Maffeo Barberini who would later be known as Pope Urban VIII and become infamous for his participation in the official condemnation of Galileo.

In the second chapter of *Science and Imagination* ("The 'new astronomy' and English imagination," first published in 1935) Nicolson offers a virtual inventory of early modern poets and writers who responded to Galileo's book; these included (in Nicolson's non-chronological ordering) William Drummond of Hawthornden, Sir Thomas Browne, Robert Burton, Phineas Fletcher, Abraham Cowley, Samuel Butler, William Davenant, Henry Vaughn, John Dryden, and Ben Jonson.[11] But Nicolson's greatest attention and praise are lavished on John Donne: "Among English poets," she writes, "none showed a more immediate response to the new discoveries than John Donne, nor is there a more remarkable example of the effect of the *Sidereus Nuncius*" (p. 46). For Nicolson, Kepler's star of 1604 and Galileo's book mark two important shifts in Donne's poetic concerns: before 1604, Donne's stars are merely conventional ("The *Songs and Sonets*, the majority of which were written before the turn of the century, contain no significant astronomical figures of speech" [p. 47]), while those that appear in the poetry after 1604—such as those that do in "To the Duchess of Huntingdon"—are important precisely because they are new:

> Who vagrant transitory Comets sees,
> Wonders, because they are rare: But a new starre
> Whose motion with the firmament agrees,
> Is miracle, for there no new things are.[12]

But the *nova* of 1604 by its very nature was transitory and eventually faded from view; likewise Donne's interest in the new stars, Nicolson argues, until 1610 when Galileo's *Sidereus Nuncius* re-energizes Donne's interest not only in the fact of the new stars, but in their meaning. The rest of Nicolson's chapter is devoted to tracing something of an evolution in Donne, from his satiric treatment of the new astronomy in *Ignatius His Conclave* through to his much more serious—and anxious—confrontation with it in "An Anatomy of the World: The First Anniversary," and other poems:

> [T]he new stars appear and fade again, but they have ceased to be mere figures of speech, and have taken in new meaning, as Donne sees the relation to cosmic philosophy. They are a symbol of the "Disproportion" and the "Mutability" in the universe of which Donne has become compellingly aware. [p. 52]

For Nicolson, Donne's efforts to negotiate the new stars is epitomized in "The First Anniversary" and as such the poem emerges as the great marker of a whole generation's crisis of knowledge—"Donne's most quoted lines, in which he reflects

---

[11] Nicolson's catalog is interrupted, as it were, with two oddities: first, a brief discussion of Christopher Marlowe, "whose imagination would have responded most sensitively to the poetic implications of the 'new astronomy' [but] died too early to know them" (1956: 41). The second is another brief discussion of a dramatist, this time Shakespeare who, though he indeed lived to learn of the new astronomy, evidently did not care much about it—though, Nicolson offers, if he "ever expressed himself on the new cosmology, he should have done so in *King Lear*, written while men's minds were dwelling on the significance of the new star of 1604" (1956: 43).
[12] John Donne, "To the Countesse of Huntingdon," [qtd in Nicolson, (1956: 48)]; for the entire text of the poem, see Donne (1990: 67–70).

the poignant regret of a generation which had inherited from the past centuries conceptions of *order, proportion, unity*, which had felt the assurance of the immutable heavens of Aristotle, take on new meaning when one reads them, remembering the revolution in thought that was occurring in 1610" (p. 52).[13]

As this brief outline suggests, for Nicolson—and, indeed, for first-wave science and Renaissance literature critics in general—reading the poetry (and less often, the prose) of a figure such as Donne is largely a matter of reading these texts "against the scientific background of his time, against the inruption of new stars and the dramatic discoveries of Galileo's telescope" (Nicolson, 57).[14] Indeed, it will take something like a virtual revolution in the study of the culture[s] of science to dislodge the influence model and to clear the way for a new understanding of the relation between early modern science and literature that is symmetrical in nature.

II

So long as the historical epistemologists—such as Singer—defined how we understood science, the influence model of literary studies was not only secure, but indeed the dominant model for understanding the relationship between literature and science precisely because it seemed the inevitable model. But even as science cannot remain science if it is static (one of the axioms of the historical epistemologists themselves), the history of science cannot remain vital if it is static. And indeed, as many scholars have noted, the history of science—as a discipline (though, as we shall see, even this designation is open to serious debate)—undergoes not only challenges, but something very like a crisis, beginning in the 1970s and that is in many ways still unfolding today. In their introduction to *Reappraisals of the Scientific Revolution*, Robert S. Westman and David C. Lindberg offer a compelling sketch of this crisis. For Westman and Lindberg, the historical epistemologist trajectory that in many ways officially begins with Herbert Butterfield's 1949 book *The Origins of Modern Science* reaches something of a culmination in Thomas S. Kuhn's landmark 1962 study, *The Structure of Scientific Revolutions*.[15] In the years since these two books appeared, and even as they have "become part of

---

[13] Interestingly, Nicolson also sees this poem and this struggle as pivotal in Donne's life itself—and an answer to the perhaps by-now tired question of Donne's conversion:

"Paradox and Probleme" Donne remains to his modern critics, who will probably never agree about the "conversion" that transformed Jack Donne into Dr. John Donne, Dean of St. Paul's . . . I shall continue to believe that the discoveries of the new astronomy, coinciding with a troubled period in his own personal life and in his age, proved that straw that broke the back of his youthful scepticism and led John Donne "from the mistresse of my youth, Poesy, to the wife of mine age, Divinity." (1956: 57)

[14] Donne's poetry and prose were of abiding interest to critics of early modern science and literature. The bibliographies of modern Donne criticism assembled by Roberts list dozens of entries; see Roberts (1973–1982) and (2004). See also Coffin (1937) and Empson (1993), especially Ch. 2, "Donne the space man," pp. 78–128.

[15] Lindberg and Westman (1990: xvii–xviii).

a canon of pedagogical texts that appear annually in history of science course lists" (p. xviii), Westman and Lindberg argue that "a generation of scholars has been whittling away at all aspects of the historiographic corpus that Kuhn's *Structure* and its sources presupposed" (p. xviii). Among the factors contributing to this change, Westman and Lindberg identify the following: the "discovery" of archives—either of lesser-known scientific figures or new facts about more prominent ones; and the development of specializations within the broader field of the history of science.[16] At the same time—and as an important dimension indicative of "the range and depth of this shift in historiographic sensibility" (p. xix)—writers and critics offered "an even more serious kind of challenge": the move from "intellectualist traditions and toward contextualization of problems and solutions in specific intellectual polities" (p. xix). This shift toward the investigation of the socially embedded nature of the scientific enterprise is clear in a work such as Steven Shapin and Simon Schaffer's important study of Robert Boyle and emergent experimental practices in early modern, *Leviathan and the Air-Pump*. This book, Westman and Lindberg argue, problematizes Boyle's air-pump experiments by "denying customary distinctions among text, instrumentation, and experimental facts and by viewing natural order not as something to be 'discovered' but as a site of meanings produced by rival interest groups in struggles over political order" (p. xix).[17] In addition to these forces shaping the history of science, Westman and Linberg also point to work in the sociology of scientific knowledge, as well as the often-cited "'linguistic turn' in cultural and intellectual history and literary theory" (p. xxi).[18] Given the rapid and extensive expansion of the kinds of study now undertaken by writers on the history of science, Westman and Lindberg assert that "historians of science are in greater disagreement today about how to conduct their craft than the ubiquitous metaphor of Scientific Revolution suggests" (p. xx).

While this comment is itself embedded within a reappraisal of the very notion of a scientific revolution, it also can be said to frame a larger discussion about the nature of the history of science as a discipline. In his own essay in *Reappraisals*, entitled "Conceptions of the scientific revolution from Bacon to Butterfield: A preliminary sketch," Lindberg offers an even more provocative assessment of contemporary "crosscurrents" within the history of science that serve, in effect, to define disciplinary challenges to the field. "Deeply affected" by a wide range of concerns—"by increasing specialization; by the methodology and content of adjoining disciplines, by alternative visions of the world; by fears about the impact of science on contemporary society or about the separation of science from the humanities"—today's

---

[16] "History of science," Westman and Lindberg write, "was beginning to be transformed into an encampment of specialists. Specialization brought with it an impatience with conceptual vignettes and broadly brushstroked stories; the new historians focused instead on 'aspects' or 'periods' of intellectual evolution, on 'discovery,' and especially on elements previously marginalized by too exclusive attention to the Greats" (1990: xviii).

[17] See Shapin and Schaffer (1985). See Shapin (1994).

[18] For two discussions of this "turn," see Latour (1992) and Lenoir (1994). Both of these are reprinted (Latour's in an abridged form) in Mario Biagioli (2003: 276–89 and 290–301, respectively).

Cultural historians of various methodological persuasions have sought to understand the relationship between science and innumerable other features of sixteenth- and seventeenth-century European culture, such as magic, religion, education, art, literature, and technology. And, in what is undoubtedly the strongest of the crosscurrents, social historians, influenced by Marxist thought or by developments in sociology, anthropology, and linguistics, have refused to treat science as a purely intellectual quest for truth and have begun to raise questions about its social and political construction.[19]

For Michel Serres, writing in his introduction to a collection of essays, *A History of Scientific Thought: Elements of a history of science*, Lindberg's "crosscurrents" in fact constitute a new model for the history of science, "a true history of science, envisaged as an autonomous discipline with its choices, its intentions, its divisions and its own style and methods." Describing this new vision of the history of science, Serres declares, "It aims to be more of an entity in itself than the deceptively clear account given in a complete encyclopedia of science covering the whole of history." Serres continues,

> Far from tracing a linear development of continuous and cumulative knowledge or a sequence of sudden turning-points, discoveries, inventions and revolutions plunging a suddenly outmoded past instantly into oblivion, the history of science runs backwards and forwards over a complex network of paths which overlap and cross, forming nodes, peaks and crossroads, interchanges which bifurcate into two or several routes. A multiplicity of different times, different disciplines, conceptions of science, groups, institutions, capitals, people in agreement or in conflict, machines and objects, predictions and unforeseen dangers, form together a shifting fabric which represents faithfully the complex history of science.[20]

A proper response to the properly historical question about the causes of the sea-change in the nature of the history (histories) of science that has led us from the historical epistemologists to the exuberant pan-disciplinarity celebrated by Serres would require a different kind of study than the one undertaken in this chapter. But in addition to the emergence of Westman and Lindberg's "shift in historiographic sensibility" and Lindberg's "crosscurrents"—specialization, contextualization, the sociology of knowledge, the linguistic or semiotic turn, etc.—I would point to a set of equally important cultural forces that have helped lead us to Serres's "shifting fabric." Among these would be the iconoclastic work of Paul Feyerabend, especially his landmark 1975 book, *Against Method*, in which he offers a radical rereading of science—in this particular case it is the work of Galileo—as marked by the rise of rationalism.[21] To this I would also add the rise of cultural studies; the emergence

---

[19] Lindberg (1990: 19).

[20] Serres (1995a: 6).

[21] In a striking passage on Galileo's astronomical writings—and on the so-called rationalism of the scientific method, more generally—Feyerabend identifies the fundamental insufficiency of rational argument:

> It is clear that allegiance to the new ideas will have to be brought about by means other than arguments. It will have to be brought about *by irrational means* such as propaganda, emotion, *ad hoc* hypotheses, and appeal to prejudices of all kinds. We need these "irrational means" in order to

of the social studies of science, particularly as represented in the work of the Edinburgh school; and the sustained challenge to science studies represented in the work of Bruno Latour.[22] Perhaps the most significant force not identified in such inventories as Lindberg offers is the crucially important and enabling impact of feminism in general, and feminist science studies in particular. Among the many figures whose work has been transformative, three of the most important are Evelyn Fox Keller, especially her critique of the Baconian ideology of the domination of nature through science; Donna Haraway, especially her work on "situated knowl-edges," and cyborg and cyborg culture; and N. Katherine Hayles and her work on posthumanism.[23]

III

In light of these new concerns about the nature of science, the nature of culture, and the nature of nature, the ways in which critics read and discuss early modern literature and its relation to science have been transformed. As suggested near the outset of this Introduction, the foreground/background model of first-wave science and literature studies has given way to a second wave comprising a more complex understanding in which not only the scientific, to the extent that it impacts or influences the literary, can be said to produce the literary, but the equally complex and important ways in which the literary, for its part, can be said to produce the scientific. Indeed, as many particular instances from the period can demonstrate, the separation between science and literature—that selectively permeable barrier that conventionally only admits "influence" to pass through from one side to the other, from science to literature—is itself more of a consequence of the emergence of science than its cause. In other words—and I take this to be a goal of current studies of early modern literature and science culture—the separation between science and literature (or, more broadly, culture) should be understood as one of the "products" of science and as such it is an ideal object of study rather than an unquestioned feature of history.

Another goal, then, of our contemporary studies is the telling of new narratives about the unity of culture-science that has always been in place but has always been obscured by the ideological separation of science from literature, and the history of the construction of the separation between science and the rest of culture. Thus, to look briefly at an example from Fancis Bacon's contribution to the Gray's Inn Revels, 1594–5 (a work that will be taken up at length in Chapter 2, below), the four "principal Works and Monuments" that he suggests should be founded and nourished by the state—the library, the garden, the cabinet of wonders, and the

---

uphold what is nothing but a blind faith until we have found the auxiliary sciences, the facts, the arguments that turn the faith into sound "knowledge" (1975: 153–4).

[22] For an example of the work in science studies identified by the Edinburgh School label, see Bloor (1976).

[23] Keller (1985), especially Ch. 2, "Baconian science: The arts of mastery and obedience." See also Haraway (1998) and (1991) and Hayles (1999).

laboratory—begin to take on a new fullness within a more rich and complex context provided by these new critical, interpretive, and theoretical practices.[24] The "collecting of a most perfect and general Library" (Bacon, 47), for example, opens on to studies of the connected natures and histories of the book (especially the illustrated book) and early modern scientific practice and discourse; or, on to the study of the evolution of new reading practices coincident with the evolution of scientific optical instruments such as the microscope and the telescope.[25] Bacon's "spacious, wonderful Garden" (Bacon, 47) emerges as the prototype of a particular organization of nature deployed semiotically and as part of an epistemological (and political) discourse borne of new understandings of the nature of nature.[26] Similarly, Bacon's "goodly huge Cabinet" and his "still-house" (Bacon, 47–8) engender new histories of "cabinets of curiosity" and the museum, on the one hand, and full-scale narratives of the socially embedded and determined space of the early modern laboratory, on the other.[27] And of course, the list could be expanded to include other sites of emergence, including travel writing, New World exploration and colonization, the discourses of "monsters," the revision and expansion of university curricula, and the early modern London theaters, among others.[28]

As this very brief outline suggests, the range of work undertaken by critics today studying early modern science and literature is vast and various. It ranges (for example) from Claire Preston's *Thomas Browne and the Writing of Early Modern Science* to Angus Fletcher's expansive study, *Time, Space, and Motion in the Age of Shakespeare* in which the critical focus is on not only "crossovers between scientific and literary expression," but, as Fletcher writes, on "conceptions underlying the most abstruse metaphysical thoughts—conceptions of the natural that should never be excluded from serious discussions of belief and hence from the workings of imaginative literature as well."[29] It includes new studies of machines and early modern technologies in Jonathan Sawday's *Engines of the Imagination: renaissance culture and the rise of the machine* and Jessica Wolfe's *Humanism, Machinery and Renaissance Literature*. It includes the subtle and powerful analysis of the relationship between Shakespeare and genetics offered in Henry S. Turner's book *Shakespeare's Double Helix*, a critical move unthinkable within the earlier model of literature and science studies. Turner proposes to advance "a simple but counterintuitive argument: that we should regard genetic engineering and biotechnology

[24] Bacon's text can be found in *Gesta Grayorum: Or, The History of the High and Mighty Prince Henry Prince of Purpoole, Anno Domini 1594*, ed. Bland (1968: 44–56).

[25] See Frasca-Spada and Jardine (2000), for example, and for a recent collection of essays on the nature and role of authorship within the cultures of science, see Biagioli and Galison (2003).

[26] See, for example, Watson (2006) and Bushnell (2003).

[27] For cabinets and museums, see Findlen (1994); Preston (2000: 170–83), Swann (2001), and Impey and MacGregor (1985). For discussions of the early modern scientific laboratory and its practices, see Shapin (1994).

[28] For discussions of travel and the New World, see, for example, Campbell (1999); Fuller (1995), and Albanese (1996). For a discussion of monsters, see Daston and Park (1998). For an analysis of the disciplines and curricula of geography, see Cormack (1997). And for discussions of the early modern theaters, see, for example, Turner (2006), Mazzio in Howes (2005: 85–105), Smith (2004: 147–68), Paster (2004), and Traub (2002).

[29] Fletcher (2007: 11). See also Preston (2005).

not simply as a new application of scientific knowledge but rather as a new mode of poetics, and that Shakespeare's own work provides a model for just such an approach."[30]

Given the richness of these new studies (and I have offered only the broadest of indications here of their range and depth) it is clear that no single example of this work can be representative.[31] But in order to highlight a bit more explicitly the innovativeness—and, therefore, the power—of this new work, I would like to mention briefly one recent book that can certainly be said to be illustrative of some of the major features of this new wave of criticism. Elizabeth Spiller's study *Science, Reading and Renaissance Literature* is dedicated, in its broadest terms, to the investigation of "a shared aesthetics of knowledge" that underwrites both scientific and imaginative writing in the period.[32] Spiller begins by clearly marking the distinction between the nature of her critical project and the more conventional arguments about early modern science and literature within what has been identified in this discussion as the influence model. Naming in her introduction the works of those scientific writers she will discuss (including William Gilbert, Galileo, William Harvey, Johannes Kepler, and Robert Hooke) and those writers of imaginative fiction she will consider (Philip Sidney, Edmund Spenser, and Margaret Cavendish), Spiller writes:

> What these texts demonstrate is that early modern science is practiced as an art and, at the same time, that imaginative literature provides a form for producing knowledge. Within this framework, literary texts become more than just topical commentaries on new scientific discoveries or intellectually (but not truly scientifically) interesting examples of the cultural work that literature might produce in the face of changing scientific knowledge. It is not just that fiction serves as a (more or less accurate) record of, as John Donne puts it, how the "new philosophy calls all in doubt." Rather, literary texts gain substance and intelligibility by being considered as instances of early modern knowledge production. Early modern fiction needs to be looked at as more than just a kind of repository for new facts or errors. [p. 2]

Similarly, science does not bear merely a superficial relation to imaginative writing—the deployment of certain "literary devices," for instance, "or narrative and rhetorical forms" effectively borrowed from the literary domain. Indeed, the works of early modern science "do not align themselves with early modern poetry because of the ways in which they are written." For Spiller—and, by extension, for the work I have called here second-wave criticism of early modern science and literature (and which could perhaps also be called the "discursive model")—the relation in question is a far more significant one in large part because the line separating science from fiction, or science from culture, is never as absolute as the historical and conventional wisdom would have us believe:

---

[30] Turner (2007: 7).
[31] In the notes to her valuable Introduction, Mazzio (2009: 13–23) offers an extended bibliographical inventory of new science studies-inflected criticism of early modern literature and culture.
[32] Spiller (2004: 3).

[S]cience maintains strong affiliations with poetic fictions because, in ways that are rarely acknowledged, its practice emerges out of a central understanding of art as a basis for producing knowledge. A belief in the made rather than the found character of early modern knowledge unites poets and natural scientists. [p. 2][33]

Spiller's study—informed by an understanding of the central role played by artifice within both scientific and imaginative writing—is characteristic of the best of the new wave of criticism: it reads scientific texts on the same playing field as literary texts; it refuses the easy separation of the "two cultures" described in C. P. Snow's classic account; and, most helpfully, it highlights the common goal shared by both early modern science and literature "[to] convert accounts of personal experience into new stories of universal truth" (p. 15).[34] It is this notion of the transformation of personal experience into something like a universally valid truth that lies at the very heart of what I will call subjective writing. One of the asymmetries that follows from the traditional notion of a separate science always in the process of perfecting its own purely rational discourses—of science, that is, which is increasingly "objectively true"—is the casual unwillingness to think of scientific writing as subjective, since the subjective (so the argument goes) is the privileged domain and, more to the point, the privileged *writing* of the literary. But new science studies in general, and early modern science studies perhaps in particular, is equally unwilling to accept this asymmetry.

IV

Drawing upon much of this work as well as a number of critical and theoretical perspectives in the fields of literary studies and science studies more generally, *The Machine in the Text* rejects the very idea of the emergence of science as the inevitable and natural consequence of the spread of rationalism and the gradual eradication of a vast and heterogeneous set of non-rational practices ranging from common superstition, occultism, magic, and the mysteries of hermeticism, on the one hand, to the mystifications of philosophical scholasticism and the dogmas of conventional religious ideologies, on the other. Predicated upon the belief that science cannot be construed as an essentially *autonomous* feature within early modern culture, this book will argue that what we take today as early modern science is more properly understood as a set of desires widely dispersed among disparate cultural practices and institutions in early modern Europe: the theater, for example, or certain meditative attempts to understand the effects of progressive disease on one's own body; the radical transformation of the cosmological order initiated by Copernican theory and expanded in the mathematical and telescopic astronomy of Kepler and Galileo; and the construction of the garden as the material

---

[33] Spiller's study ranges over a number of scientific fields, including anatomy, embryology, experimental philosophy, magnetism, astronomical observation, optics, and microscopy.
[34] Snow (1959).

manifestation of a certain philosophical disposition toward abstraction and non-materiality.

These practices and institutions share between them a number of defining concerns that it will be the goal of this study to describe in detail—including especially the concern with the systematic reproduction of the event under consideration in the model of artificial experience. In the physical sciences particularly (though not exclusively, as my chapters on Shakespeare's theater and Donne's meditative prose will insist), this desire takes the form of the experiment: the controlled (or contrived) reproduction of physical events. Each of the discourses I examine in this project has its corresponding desire and its corresponding mechanism, all of which can be called artificial in both nature and function. In *Hamlet*, for instance (and Shakespearean theater more generally), artificial experience emerges from the self-consciously literary uses of such systems of representation as enclosed plays, posthumous life stories, and secret forgeries.

Similarly, in texts by John Donne, the body—one of the principal sites for early modern scientific speculation—emerges as itself a prosthetic mechanism for objectification in the service of various, and sometimes competing, definitions of experience and meaning. These issues are most pressingly at stake in Donne's *Devotions upon Emergent Occasions*, perhaps the first instance of what we eventually come to recognize as the scientific autobiography. For Galileo, whose twofold project is to invent the instruments and experiments of astronomical observation as well as the literary and graphic forms of their communication, the culture of visual knowledge comes to characterize his work as scientist and writer. But how exactly does the artful and artificial representation of experience function? If celestial objects such as our moon or the moons of Jupiter are separated from us not only by virtue of their great distances from earth, but also by reason of our limited natural sight of their features, what then are the politics of any second-order representations that offer as natural facts those celestial features that are themselves only "real" for earth-bound observers as a consequence of the prosthetic use of the telescope—both as an observational tool and as a writing instrument?

In the horticultural and garden theory of John Evelyn (seventeenth-century English virtuoso, diarist, and founding member of the Royal Society), the prosthetic objects of his garden practice turn out to be the very objects of nature whose status as natural can be established only by virtue of their artificial placement within a contrived physical setting dedicated to the artificial experience of "nature." Evelyn's garden theory can be said to represent a culminating moment of the interrelated practices of early modern literature, philosophy, science, aesthetic theory (whether literary, architectural, or art-historical), and representational poetics.

The Machine in the Text traces the emergence of a general cultural will to construe the material world not merely as a mechanical system, but more profoundly still as itself the artifact of human action. As suggested above, the various discourses and practices I address in each chapter depend upon a particular set of maneuvers dedicated to the invention of particular machines that, in turn, serve the production of certain forms of knowledge and certain types of experiences. In some instance, these are literal machines (Galileo's telescope, for instance), while others are machines to the extent that they are assemblages of instruments, strategies, and frequently technical operations: the Shakespearean stage, for example, upon which we see enacted Hamlet's efforts to reconstitute knowledge and meaning through theatrical illusion, or upon which the statue of Hermione is so obscurely animated before our very eyes. In other instances, these machines are the amalgamation of discursive practices and literal pieces of the material world: John Donne's invention of scientific autobiography fashioned from the very symptoms of disease, or John Evelyn's gardens that work simultaneously as representation (of nature and its objects) and at the same time as the radical reduction of nature into a wholly artificial semiotic.

The title of this book—*The Machine in the Text*—is also meant as something of an echo of two famous and important phrases, one of which is the title of Leo Marx's study of the fate of pastoralism in the American nineteenth century, *The Machine in the Garden*, and the other of which comes to us from the British philosopher Gilbert Ryle whose theory of mind yields the well-known devastating critique of Cartesian dualism that is summed up in the phrase, "the ghost in the machine."[35] In both instances, these phrases stand some distance behind *The Machine in the Text*. In the case of Marx, I certainly have been influenced by his entire discussion of the pastoral–industrial (or, pastoral–technological) agon he charts through his study; but I feel an even greater debt to his more theoretical concerns with the relationship between literature and history that lies behind or is figured in the intrusion of the machine into the garden—what Marx calls "the relation between literature and that flow of unique, irreversible events called history" (p. 28). At the same time, Marx's discussion of *The Tempest* as Shakespeare's "American fable" has been very influential in our readings of Shakespeare's New World play and it has influenced my thinking about *The Winter's Tale* and the early modern theater in general.

In the case of Ryle, his critique of Descartes and his theorizations of the *cogito* and the idea of thinking without the body that attends so intimately upon it (matters I address in Chapter 5) have given a powerful vocabulary for a more general discussion of the relation between mind and the body that it in some perhaps obscure way houses it. At the same time, Ryle's ghost offers a more general perspective on the idealist–materialist debate that has been central to certain forms of early modern writing, whether philosophical in nature (such as Descartes), or meditative (such as John Donne, as I argue below).

As the preceding paragraph suggests, even as the primary focus of this book is the early modern period and those writers, both literary and scientific, whose works are addressed in the following chapters, another major objective of this book is to introduce a certain—and critically generative—interplay between these early modern texts and a series of works by mid-twentieth-century writers and theorists, including (among others), T. S. Eliot, Walter Benjamin, C. P. Snow, William

---

[35] Marx (1964); Ryle (1949).

Empson, W. K. Wimsatt, Clifford Geertz, and Roland Barthes. My concern here is not strictly to bring the twentieth-century writers and their works to bear on early modern texts and practices as analytical tools meant in some fashion to decipher otherwise obscure or inscrutable aspects of seventeenth-century culture. Rather, it is to allow the early modern texts to facilitate a reassessment of aspects of the intellectual and cultural history represented by the twentieth-century writers.

The Machine in the Text refers to more contemporary discussions and theorizations of the machine within the context of the study of science. In this regard, the notion of the machine that informs this study follows directly on the work of Andrew Pickering, for whom the machine is dedicated to the capture of non-human agency, and Bruno Latour, for whom the machine is best understood as "machination."[36] In both instances, the critical matter at hand is the re-theorization of the machine, particularly in its relation to agency. At the same time, this machine—the machine that comes to us from contemporary science studies, but that is also on display in early modernity—this agential machine is itself a particular response to the problem of the relationship between nature and art, between realism and constructivism. As such, the machines I wish to study here achieve a number of objectives, the most important one of which is to trouble our otherwise too-settled understanding of the relationship between fact and artifact.

One word, as well, about the subtitle: *Science and Literature in the Age of Shakespeare and Galileo*. In addition to putting into play at least four of the principal agents discussed in this book, this subtitle is also offered as a chiastic figuration, pairing science with Shakespeare, and literature with Galileo as a way of marking at the outset my abiding interest in crossing these disciplinary categories. I want to mark this interest for a number of reasons; not least is the belief that the solidification of science into Science and literature into Literature, and the corresponding solidifications of Science and Galileo, and Literature and Shakespeare as seemingly self-evident ontologies, stand among the primary targets that this book seeks to contest.

I begin *The Machine in the Text*, and the book's project of arguing for a new understanding of the natures of and relationship between early modern science and literature more generally, by examining the textual record of the Christmas-season festivities elaborately staged by one of London's Inns of Court near the end of the sixteenth century, the *Gesta Grayorum*. In Chapter 2—"Gray's Inn Revels, 1594–5"—I argue that the Revels themselves, together with certain critical readings of them, provide the opportunity for an interrogation of the nature of what I will call disciplinary knowledges. My concern here will be to demonstrate the ways in which the meanings of objects of study are, to a significant degree, determined by the disciplinary practices brought to bear upon them. Thus, for example, the *Gesta Grayorum* comes to have a certain meaning for Shakespeareans for the information it contains concerning the first performance of *The Comedy of Errors*—the most complete and comprehensive record, in fact, that we have of the first performance

---

[36] Pickering (1995), Latour (1987).

of any play by Shakespeare. For other literary critics and historians, because the Revels served as the setting for Francis Davison's *Masque of Proteus*, the *Gesta Grayorum* stands as an important document for our understanding of the court masquing practices that will reach their full flowering in the hands of Ben Jonson and Inigo Jones in the early seventeenth century. At the same time, for other readers—social historians, for instance, or historians of early modern science—the *Gesta Grayorum* is an important document for what it reveals about the inner-workings of the Inns of Court as a training ground for the governing class, or for what light it sheds on a relatively obscure text by a major figure in the history of early modern science: Francis Bacon's *Device*, an early articulation of Bacon's programmatic vision for the reformation of natural philosophy.

However, having once demonstrated some of the ways in which the Revels help us to understand the emergence of disciplinary knowledges, I then trouble this too-easy narrative by demonstrating how artificial a neat division into literary and scientific meaningfulness in fact is—and by doing so I seek to question the status of strictly disciplinary knowledge and disciplinary practice. The *Gesta Grayorum* enables a new understanding of the mutually sustaining nature of these two discourses. Indeed, by in effect submitting the literary text to the methods of natural philosophy and the natural philosophical to the methods of literary analysis, I hope to demonstrate just how thoroughly imbricated are these two discourses.

Chapter 3—"Hamlet's Machine"—considers the question of experience on the Shakespearean stage, and in the Shakespearean world more generally. The point of departure for the work of this chapter is what can be called the critique of the loss of experience in modernity—a critique, I will argue, that is sensed, though largely misunderstood, in T. S. Eliot's famous (or perhaps infamous) essay on *Hamlet*, "Hamlet and his problems," and that is the object of the sustained and powerful philosophical analysis in the work of Walter Benjamin. In *Hamlet*—that text (or that "theory," as it were) that at the opening of the modern world in many ways inaugurates the crisis felt in Eliot's essay and interrogated in Benjamin's work—experience is destabilized or displaced through the collapse of the perceptual body, beginning with the destruction of the "smooth body" of Old Hamlet and continuing throughout the rest of the play, even into the grave and Hamlet's sepulchral curiosity about the dissolution of the body in the earth.[37]

In the aftermath of the collapse of the perceptual body—and the traditional epistemology constructed upon the proper functioning of the senses and the mind's ability to fashion knowledge based on sense perception—Hamlet is left to discover or fashion an alternative method for the recuperation of knowledge and action in the world. Hamlet's response—Hamlet's hope—is to attempt the recovery of knowledge through a strategic consolidation of experience derived from perception re-deployed within a network of practices and techniques that together render experience artificial and evidential. Hamlet secures a way to knowing through those

---

[37] Shakespeare (1982: 1.5.73)

practices that serve to construct what will become the defining feature of science, the experiment.

The production of "natural artifacts" is the focus of Chapter 4, "Galileo's Telescope." This chapter discusses one of the great revolutions in early modern cosmology: the discovery in the first decade of the seventeenth century of irregular and apparently migratory blemishes on what was long believed to be the pristine surface of the sun. In his *Letters on Sunspots* (1613), Galileo understood these spots to constitute definitive proof against the Aristotelian belief in a materially perfect universe: The sun's "supposed immaculacy," Galileo wrote, "must yield to obser-vation."[38] Rejecting assertions that the spots were merely clouds above the solar surface or simply flaws in the lenses of his telescopes, Galileo insists the sunspots are as real as dark spots on the moon produced by the shadows cast by lunar mountains (another Galilean discovery)—and moreover, although the spots, too, are bright, "in comparison with the lighted portions they are as dark as is the ink with respect to this paper" (p. 93). While for Galileo sunspots come to stand as proof for the sphericity and the rotation of the sun, and, finally, the sun's stationary position within the solar system, I argue in this chapter that the primary existence that could be claimed for the sunspots was indeed a matter of "ink with respect to . . . paper." Using the invention of sunspots as a case study, this chapter presents a counter-narrative to the conventional understanding of the relation between vision and belief that is said to underwrite the so-called scientific method. Rather than serving as an instance of sight conferring belief, the invention of sunspots demonstrates the ways in which belief is first required in order that there could be a confirming vision.

The detection of sunspots was enabled by the adaptation of the terrestrial telescope to celestial observation. But within this set of technical practices, sunspots are made "real" only by the very technology (or mechanics) that enabled observa-tion in the first place: projecting what Galileo called the "luminous cone of sunlight that emerges from the telescope" (p. 116) against a writing surface and tracing on it the solar disk and shading in the sunspots he "saw" on the "sun." Sunspots were invented by the telescope and the technologies of representation deployed in order to make them manifest both to the observer and—through the technologies of print—to his readers. Sunspots are made "real," in other words, through the use of the telescope as both a tool of observation and an instrument of writing.

Although Galileo succeeded in both establishing the existence of sunspots in general and the yet more remarkable feat of determining their meaning in particular (ultimately, the validity of the Copernican world system—and the place in it for Galilean mechanics), he was just as famously less successful in avoiding charges of heresy leveled against him by a Church made nervous, to say the least, by the emergence of the new science and that new figure of the scientist/"natural philoso-pher" who was thought to threaten belief. Although Galileo himself evidently felt no particular conflict between pure faith in a Christian God and belief in science,

---

[38] Galileo, *Letters on Sunspots*, in Drake (1957: 135).

the "Galileo affair," as it has been called, served to make explicit the conflict between religion and science. In Chapter 5—"John Donne's New Science Writing"—I offer a discussion of the religious faith and scientific belief controversy by way of an analysis of a major text by a confirmed believer who doubted science, John Donne's 1624 meditation on disease, the body, and the problem of embodiment, *Devotions Upon Emergent Occasions*. In Galilean science, knowledge was "exteriorized" in the sense that it was knowledge about the natural world "out there," independent of any particular observer, and was therefore offered as *objective* knowledge (even if today we would want to trouble the very notion of purely objective knowledge). By contrast, the knowledge Donne constructs through his *Devotions* is rigorously "interiorized." The questions that lie at the heart of this discussion are these: How can Donne translate an interiorized—or a subjective— knowledge into a universal truth? What particular *method* can Donne deploy in pursuit of this goal? What is the theology of exemplarity? What role does autobiographical writing play? And, for Donne as writer, what are its poetics?

In response to these questions, I will argue that Donne's disease and his textual response to it provide him with the opportunity to investigate the manner of God's method—of communication, mercy, and ultimately redemption—that will serve as the model for his own critical reaction to both the epistemological and experiential crises of disease and doubt. In the face of these issues, Donne must fashion a new method—of observation, analysis, and recording. Through the meditative and autobiographical work of the *Devotions*, Donne constructs a discourse, which I will call the *new science writing*, that enables the proper apprehension of God's method and that allows for a new understanding of embodiment and materialism, together with a new model for human experience.

Chapter 6—"Nature's Art"—addresses two responses to the ancient debate over the art–nature relationship, responses that reflect different epistemologies evident both in and between literary and scientific discourses in the early modern period. First, I begin this analysis with a discussion of Shakespeare's late play, *The Winter's Tale*, a play in which we see Shakespeare negotiating the art–nature contest in two separate but complexly—but perhaps antithetically—related moments. The first is the declaration offered by Polixenes in the play's famous pastoral scene, and within the context of his natural philosophical debate with Perdita about grafting, that "art itself is nature." The second is the conjuring of Hermione's statue in the final scene of the play, a spectacular event that depends upon the logic of the *trompe l'oeil* figured not only in Giulio Romano's statue, but in play itself, that serves to contest Polixenes's naturalness of art argument.

The second part of the chapter shifts from the pastoral fantasies of *The Winter's Tale* to the world of the early modern garden and, in particular, the response to the art–nature debate that emerges from within the rigorously scientific mid-seventeenth-century garden theory and practice. In this part of the chapter I focus my analysis on John Evelyn's *Elysium Britannicum* (together with other important and related horticultural and literary texts) and argue that the early modern garden is an ideal locus for a consideration of the "*Natural* in nature" that Roland Barthes describes as the "frisson of an enormous machine which is humanity tirelessly undertaking the create meaning."[39] As Evelyn's garden writings suggest, the prime meaning produced by this machine dismantles the naturalness of art argument and replaces it with the understanding of the *artificial* nature of nature.

I locate the emergence of the garden and artificial nature in relation to the waning of what I call the "emblematical" epistemology and its replacement by the scientific. Beginning in the middle of the seventeenth century, and in the garden working and garden writing of men such as Evelyn and others associated with the founding of the Royal Society, the notion of the garden as strictly an emblem of God's inscription of absolute theological meaning in the world is gradually replaced by an understanding of the garden as the site of the human production of meaning. Among related topics and figures, I discuss the garden and the politics of representational mimesis, the rise of the georgic (as both literary form and as an epistemology) out of the seventeenth-century ruins of pastoralism, the emerging fashion of automata, the collective efforts of the Royal Society to realize Bacon's dream of institutionalized natural philosophy, and the (poetic and political) works of George Wither—all of which, I conclude, contribute to the production of the garden as the discourse par excellence of artificial nature.

In Chapter 7, "Time's Arrow," I return to a series of questions and challenges posed famously by C. P. Snow in his 1959 Rede Lecture at Cambridge University, *The Two Cultures and the Scientific Revolution.* In this chapter I will offer what I would argue has been on short supply in the seemingly endless debate over the two cultures ever since Snow's original lecture: a careful and critical reading of the arguments Snow actually offered. There certainly has been no shortage of responses to Snow's essay, the appearance of which occasioned, in fact, a small industry of reply after reply that continues even today—a history (let me add) in which F. R. Leavis's ugly polemic *Two Cultures?: The significance of C. P. Snow* holds a place of notorious prominence.[40] But where virtually all respondents tend to focus on Snow's assessment of the science–literature agon—and, more particularly, on Snow's perceived privileging of science over literature—I want instead to interrogate the assumptions about the nature of science and the nature of literature that inform Snow's discussion.[41]

In this regard, this chapter is less a contribution to the so-called Culture Wars (or Science Wars) than the attempt, on the one hand, to understand Snow's argument about cultures, and, on the other, to understand better Snow's rejection in *A Second Look* (his reconsideration of the two cultures controversies) of thermodynamics as the ideal model for intellectual inquiry, which he had proposed in *The Two Cultures*, in favor of molecular biology and the then-new science of DNA. Among the many factors that served to prompt this paradigmatic change that results in Snow's identification of molecular biology as the ideal measure of

---

[39] Barthes (1972: 219).

[40] Leavis's article was originally published in 1963 and in abridged form it, along with a great deal of related contributions to the debate (including an important essay by Lionel Trilling), is reprinted in Cornelius and Vincent (1964).

[41] Snow (1998: 22). This edition also includes Snow's subsequent essay, *The Two Cultures: A second look*, originally published in 1963.

scientific literacy and at the same time the ideal intellectual model for both cultures, is the consequences for Snow of the discovery of the double-helix: "Nature, when interested in what we call life, appears to have a taste for the rococo." Snow discovers, in other words, that at the molecular level, the material world *is* a work of art. From this perspective, we will have returned to the art–nature debate that was central to the literary and scientific works under consideration throughout this book.

In my conclusion, I offer a brief review of the territory covered in the book's six chapters: where we have been, what we have seen, and what (ideally) we have learned from the journey. I also map out the very general outlines for future work in the field of early modern studies that might continue and extend the insights represented by the many critics working toward a new understanding of the early modern cultures of science. What forms might those new contributions take? What may be their special objects of study, made visible through the wide range of methods under development today?

As I hope to establish through the work undertaken in these chapters, what has typically been alleged as an exceptionalism for science—as that practice that best accesses natural fact and truth—is instead the end result of a complex historical process through which diverse practices become consolidated under the disciplinary name *science*. It is only by way of the gradual conceptualization of a unifying culture that these practices become detached from their organic locales and implicated within the growing ideology of science. Emergent early modern science *itself* functioned as a machine for this construction and consolidation of like practices and the privileged production of natural meaning. In other words, science *understood as a culture* is the first experimental product—and production—of science.

2
Gray's Inn Revels, 1594–5

The frontispiece of the April 1966 number of the *Bulletin of the Atomic Scientists* bears a beautiful image of the x-ray diffraction pattern of beryl. This image—so precise and so precisely geometrical in character and form as to appear to trouble the line between the natural and the artificial—is accompanied by the following quotation: "The monuments of the mind survive the monuments of power." This quotation—this *axiom*, we can say, for it carries its oracular quality on its sleeve—in turn carries the following legitimating citation: "*Gesta Grayorum*, Sir Francis Bacon, 1561–1626," a citation that, by virtue of its reference to so important and foundational a figure in the history of science as Bacon is generally held to be, works to underscore the journal's general mission to speak truth to power.[1]

The *Bulletin of the Atomic Scientists* was founded some twenty years earlier in the immediate aftermath of the success (if that is the right term) of the Manhattan Project by a pair of scientists—Hyman G. Goldsmith and Eugene Rabinowitch—who had worked on the research toward and production of the first atomic weapon. These were men who had come to understand the profound complications—moral, to be sure, but also social and political in nature—that attended upon the inauguration of the nuclear age. Their journal was intended as a forum for the sustained discussions they thought necessary to the unprecedented problems and dangers that atomic weaponry had introduced to the world.[2] Fundamental to this mission was an understanding of the integral connection between science and the world of human affairs, a connection that, on the one hand, is signaled in the journal's complete title: *Bulletin of the Atomic Scientists: A journal of science and public affairs*, and that, on the other hand, had been announced with something like absolute clarity by the epochal destruction of two Japanese cities in the late summer of 1945.[3]

---

[1] *The Bulletin of the Atomic Scientists: A journal of science and public affairs* 22(4) (1966), frontispiece.

[2] The first number of the journal (originally published under the title the *Bulletin of the Atomic Scientists of Chicago*) identifies The Atomic Scientists of Chicago as an organization founded in 1945 and with a membership of 200, each of whom has signed a Declaration of Intent to uphold the aims of the group articulated in its constitution; these aims include "1) To explore, clarify and formulate the opinion and responsibilities of scientists in regard to the problems brought about by the release of nuclear energy, and 2) To educate the public to a full understanding of the scientific, technological and social problems arising from the release of nuclear energy" (1945: 1).

[3] The deliberate linking of science and public life and public affairs is not unprecedented and does not begin with the *Bulletin*. There is of course a long history, from the founding of the Royal Society forward, of attempts at popularizing science and its more dramatic or theatrical effects. Similarly, in the history of critical debates about science as a matter of public affairs there are many significant milestones that both pre-date and follow the efforts of the *Bulletin*—including, to cite only a few

As one would expect, the main contributions to the April 1966 number of the *Bulletin* support the mission of the journal to bring the work of science into the arena of public debate and include (in addition to book reviews, letters, and reports on related activities within the area of science and public affairs) an article co-written by a biophysicist and a mental-health researcher entitled "How to keep the peace in a disarmed world," for example, and an article by the then Director of the Oak Ridge National Laboratory, Alvin M. Weinberg, entitled "Science, choice, and human values," a piece that is prefaced by the following editorial statement:

> The enthusiastic and lavish support which scientific and technological research have enjoyed in America ever since Hiroshima, or at least since Sputnik, is being tempered by second thoughts. How big scientific and technological research should become—even how to frame this question properly—is now the subject of much soul-searching within the government and within the scientific community. [*Bulletin* (1966: 8)]

While this editorial preface bears the marks of its particular historical moment (even, indeed, as its central concern over the exact parameters of what today we know as Big Science seems perfectly apt in the opening years of the twenty-first century), at the same it time seems to harken back to the early modern period in which Francis Bacon and his efforts to reform natural philosophy played such a key role. The Bacon passage offered in the prologic place of honor is, in a word, a creation narrative, part and parcel of the conventional identification of Bacon as a leading figure in the emergence of what he christened "the Great Instauration"— and a matter of some serious debate (as I will discuss below).

The meaning of Bacon's assertion seems straightforward enough—at least at first glance: intellectual effort and success will prove more enduring than the effects of sheer political power. The case, though, is rather altered when one takes a second look and either tries to index such a moral to the politics of the *Bulletin*, or when one returns to Bacon's writing in order to encounter the axiom *in situ*, in its local context and in its narrative or descriptive fullness. The immediate problem that arises when trying to use the argument that the works of the mind are more permanent—and more valuable and, by implication, more *moral*—than the works of power within the context of nuclear weaponry is that the fact (let us call it) that the nuclear bomb is simultaneously an effect of both mind and power, if "power" is understood to mean the combination of political will and vast capital investment. Indeed, the bomb is an unequivocal demonstration of a particular kind of *inseparability* of mind and power enabled by the advent of an especially coherent and efficient instance of Big Science. As such, the Baconianism (if that is what it is) does not easily attach its morality to either the mind (or, the science) that created the bomb or to the power that enabled its development and subsequently wields its force (or the threat of its force).

---

landmark examples, T. H. Huxley's 1880 "Science and culture," Matthew Arnold's 1882 Rede Lecture, "Literature and science," Huxley's own Rede Lecture (1883) and—somewhat later but decidedly more famously—C. P. Snow's 1959 Rede lecture, *The Two Cultures*. Snow's lecture and the entire two-cultures concept is the subject of Chapter 6, below.

An alternative reading of the axiom is that the work that appears in the *Bulletin* is meant to be representative of mind, and the nuclear arms race inaugurated with the Manhattan Project and dutifully sustained by the two superpowers through the period of the Cold War (just heating up, in fact, in the wake of Sputnik and in the mid-1960s) are meant collectively to represent power. But even this more hopeful reading is in fact complicated by the quotation and citation heralded on the frontispiece: "The monuments of the mind survive the monuments of power. *Gesta Grayorum*, Sir Francis Bacon, 1561–1626." If by my third or fourth repetition of this axiom it begins to sound particularly modern, or particularly twentieth century, that is because (like the bomb and like the Cold War) it is in fact a twentieth-century invention. If we return to Bacon, we find that the line as quoted in the *Bulletin* is, strictly speaking, a fiction, for not only does it not appear in the *Gesta Grayorum*, but it does not appear in *any* work by Bacon. I do not mean these observations as a criticism of the work of the editors of the *Bulletin*—or the sentiment (or argument) behind the attempt to ground their political message in the work of Francis Bacon, especially since there is indeed much in Bacon's work to support just such an argument. To the contrary, I applaud that effort and, at the same time, want to suggest that the errors in quoting and citing can be fruitful and productively engaged.

Perhaps the first step in this process is to see the quotation not as either literal (as it is indeed offered) or as fictional (as I suggested a moment ago), but rather to see it as an approximation, an improvised redaction of what is taken to be a Baconianism. The line to which the frontispiece improvisation gestures appears in a text that bears much resemblance to Bacon's contributions to the *Gesta Grayorum*, a dramatic device also prepared for a courtly entertainment. In this case, it was the 1595 Essex entertainment.[4] In the Essex entertainment, the line appears in this form: "The monuments of wit survive the monuments of power: the verses of a poet endure without a syllable lost, while states and empires pass many periods" (p. 62). The *Bulletin* improvisation has made a few cuts and one emendation, substituting "mind" for the presumably less modern and therefore less clear "wit." But these acts of revision or improvisation are not to be scorned; rather, as I will discuss below, such improvisation is not only common, but in fact it constitutes the method through which the objects of our disciplinary study—Bacon, for example, or his contribution to the *Gesta Grayorum*, as another—can be said to come into being. Bacon himself will later revise and expand the line from the Essex entertainment and offer it in its new—and more powerful form—in his much more famous text, *The Advancement of Learning*, in 1605:

> We see then how far the monuments of wit and learning are more durable than the monuments of power or of the hands. For have not the verses of Homer continued

---

[4] This text appears under the title "Of love and self-love" in Bacon (1996); unless otherwise noted, all quotations from Bacon's work refer to this edition. Vickers identifies the Essex entertainment as "the last of the proto-dramatic entertainments that Bacon was involved with in the 1590s" (Bacon [1996: 535]). For a discussion of the details of the Essex entertainment, including its plot features, participants, and its bibliographic history, see Vickers's long note to the text, pp. 535–7.

twenty-five hundred years or more, without the loss of a syllable or letter; during which time infinite palaces, temples, castles, cities, have been decayed and demolished? [p. 167]

It is perfectly clear from this passage (and others like it throughout Bacon's writings) how the new world of print culture helped to structure Bacon's thinking about the nature of inquiry, method, and knowledge. The resolutely textual nature that *wit* takes in this touchstone passage in *The Advancement of Learning*, however, functions in two opposed ways. On the one hand, it serves to enshrine the textual and print nature of knowledge, and hence its having achieved something like a functional immortality (unlike the decayed and demolished palaces and cities). On the other hand, the sentiment expressed in this passage works against the very idea of origins—even in the printed texts of new Baconian knowledge themselves. Having just celebrated the stability of knowledge in its textual form, Bacon goes on to subvert the very durability of textualized knowledge in favor of a more fluid understanding of texts less as final statements of new knowledge, and more as points of departure for future improvisations—what Bacon calls "perpetual renovation":

> It is not possible to have the true pictures or statuaes of Cyrus, Alexander, Caesar, no nor of the kings or great personages of much later years; for the originals cannot last, and the copies cannot but leese of the life and truth. But the images of men's wits and knowledges remain in books, exempted from the wrong of time and capable of perpetual renovation. Neither are they fitly to be called images, because they generate still, and cast their seeds in the minds of others, provoking and causing infinite actions and opinions in succeeding ages. [pp.167–8]

This notion of "perpetual renovation" is important when we consider the afterlife of both Bacon's *Device* and the *Gesta Grayorum* more generally, with which it is very often identified. Indeed, as the following discussion will seek to establish, Bacon's contributions to the Gray's Inn Revels, 1594–95 (and the *Gesta Grayorum*, which is its textual embodiment) have caused if not "infinite" at least numerous and important critical "actions and opinions in succeeding ages."[5] More particularly, the Revels—and frequently Bacon's roles in them—have become variously interesting to diverse writers and critics from a range of disciplines.[6] In the following pages, I would like to focus on the nature of the set of disciplinary knowledges within which Bacon's part in the Revels has become important. And it follows that within these particular disciplinary regimes, what constitutes knowledge is necessarily variable.

---

[5] *Gesta Grayorum: Or, The History of the High and Mighty Prince Henry Prince of Purpoole, Anno Domini 1594*, ed. Bland (1968). Subsequent references are to this edition, unless otherwise noted, and abbreviated *GG*, followed by page numbers. As Bland describes in his Introduction (*GG*, ix–xii), the *Gesta Grayorum* was originally published, under somewhat unclear circumstances, nearly one hundred years after the Gray's Inn Revels, in 1688. It is clearly a first-hand account of the events it narrates and describes (perhaps by one of the participants in the Revels). For a discussion of the bibliographical details of the 1688 publication, see *Gesta Grayorum*, ed. Greg (1914).

[6] In addition to the literary and scientific examples to follow, I would also point to an interest found in work by scholars interested in early modern law; see, for example, Raffield (2004).

Consider the famous reference to Shakespeare in the *Gesta Grayorum*, which constitutes perhaps the most elaborate depiction of the specific setting and context for a first performance of a play by Shakespeare. The play in question is *A Comedy of Errors* and the report of what indeed appears to have been its inaugural performance is perhaps familiar to readers of Shakespeare's early play—a matter I will return to presently. "When the Ambassador was placed," the report reads, "and that there was something to be performed for the Delight of the Beholders, there arose such a disordered Tumult and Crowd upon the Stage, that there was no Opportunity to effect that which was intended" (*GG*, 31). In the face of this "Tumult," the "Ambassador and his Train," thinking "they were not so kindly entertained," walked out altogether. Even after their departure, the tumult continued to the point at which "it was thought good not to offer any thing of Account" (*GG*, 31). But all was not entirely lost and the evening continued with "Dancing and Revelling with Gentlewomen; and after much Sports, a Comedy of Errors (like to *Plautus* his *Menechmus*) was played by the Players. So that Night was begun, and continued to the end, in nothing but Confusion and Errors; whereupon, it was ever afterwards called, *The Night of Errors*" (*GG*, 32).

As this passage makes clear, Shakespeare's play is merely incidental to both the evening's entertainment (it is identified, after all, as a thing literally of no "Account") and to the *Gesta Grayorum* more generally. The same could be said, for that matter, of the "Players" who are referred to in the text as "a Company of base and common Fellows" who had been "foisted" on the Revellers in order to cause precisely the sort of "Discredit" (*GG*, 33) to the members of Gray's Inn as the "Night of Errors" indeed produced (or is said to have produced). And what may appear to today's readers of *A Comedy of Errors* as merely incidental—the presence and then departure of the Ambassador, for instance, the "Tumult" itself, or, indeed, the entire Christmas Revels—were clearly central to the account offered in *Gesta Grayorum*. In other words, from a particular perspective—one we can call "Shakespearean"—the Revels and all their variety of entertainments are relegated to an only contextual significance and, naturally enough, for readers invested in the Revels as a more or less anthropologically interesting historical account (as for the authors of and participants in the Revels themselves), *A Comedy of Errors* barely achieves notice (much less attention)—and does so only in the service of retelling details of the "Tumult" and its aftermath.[7]

These assertions may seem obvious enough and hardly worth mentioning. But in fact, the status of *A Comedy of Errors* in the *Gesta Grayorum* as conspicuously visible

---

[7] In keeping with the presiding spirit of improvisation and play, the member of Gray's Inn took the occasion of the "Night of Errors" to stage an investigation into the causes of the "Tumult" and the identity of the instigator. Though an embarrassment "to our whole State," the chaos did provide "occasion to the Lawyers of the Prince's Council, the next Night, after Revels, to read a Commission of *Oyer* and *Terminer*" to authorize the inquiry into the "Disorders and Abuses" visited upon the Prince and his court, "by Sorceries and Inchantments; and namely, of a great Witchcraft used the Night before, whereby there were great Disorders and Misdemeanours, by Hurly-burlies, Crowds, Errors, Confusions, vain Representations and Shews, to the utter Discredit of our State and Policy" (*GG*, 32). This further play allowed the pleasure of mimicking the legal procedures of judicial proceedings—the very thing for which many members of the Inn were then in training.

and by turns inconspicuously invisible serves to illustrate a fundamentally important feature of what I want to call the disciplinary nature of knowledge for which a given text or the study of a particular history (among other things) may well stand as illustrative. And the *Gesta Grayorum* is a text (even as the Revels were an event) that is especially rich ground for the interrogation of the nature of early modern disciplinary knowledges. As such, it follows, then, that the object under a given disciplinary study may well be imagined to emerge as an essentially different object than that which emerges from another disciplinary perspective and discourse. So (to continue with the present example), within the Shakespearean disciplinary regime, the *Gesta Grayorum* exists more or less as a piece of documentary evidence placed in the service of dating a play or helping to establish a chronology, among other strictly disciplinary concerns. For other readers—for historians, for example, or for historians of science—the *Gesta Grayorum* is important for a completely different set of reasons: because it helps us understand the social dynamics of the men of the Inns of Court, or because it allows us a certain and perhaps privileged glimpse into an early instance of the clear articulation of what will eventually become known as the Baconian declaration of the pursuit of natural philosophy under the auspices of the prince.

For the purposes of the present discussion, it is safe to say that the "perpetual renovations" of Bacon and the *Gesta Grayorum* fall into two broad categories: the literary (of which the Shakespearean is only one subset), and the scientific. Although I will presently pursue both of these broad disciplinary regimes in more detail separately, it is important to stress that such a separation is entirely artificial. Indeed, the virtue of this analysis of the *Gesta Grayorum* lies precisely in coming to see the very great degree to which the literary and the scientific (which now I should identify within quotation marks, thus: the "literary" and the "scientific"—or, even, "the literary" and "the scientific," to register the multiple versions available within each category) as mutually informing and mutually sustaining. Preparatory to this discussion, it is well to offer an overview of the events of the Gray's Inn Revels.

Gray's Inn was one of the four Inns of Court, along with Lincoln's Inn, Inner Temple, and the Middle Temple—institutions dedicated to the study and profession of the law. They were, in the words of Desmond Bland, the editor of the *Gesta Grayorum* cited in this chapter, "unique institutions, at once the training ground and the governing body of the barrister level of the legal profession" (*GG*, xvi).[8] It was traditional for the Inns to sponsor and undertake to produce various festivities throughout the year, with special efforts expended for the Christmas holiday season. The 1594–5 season was noteworthy for the especially elaborate (and expensive) Revels, a fact at least in part attributable to the fact that the Revels had been suspended for some three or four years prior, "by reason of Sickness and Discontinuances" (*GG*, 5). The Revels began on 20 December and continued

---

[8] Bland (1968) offers a brief but useful account of the evolution of the Inns of Court, particularly in relation to the traditions of English Common Law; see pp. xvi–xxiv. For historical studies of the Inns in relation to the theater, Bland references Green (1931), as well as his own bibliographical study of the subject (1965). See also Prest (1972).

(intermittently) through Shrovetide (3–4 March 1595) and included, among dozens of planned—and frequently *scripted*—events, the election and "enthrone-ment" of a master of ceremonies or Lord of Misrule, known as the Prince of Purpoole (in honor of the manor that housed Gray's Inn); the ceremonial enter-tainment of members of the Inner Temple, which was also the occasion of the "Night of Errors" discussed above; a mock trial of the "sorcerer" held accountable for the "Tumult"; a ceremonial progress of the Prince and his entourage through the city of London to dine at the Lord Mayor's house; an entertainment in support of the Prince's "journey" to offer military aid to the besieged Emperor of Russia in his war with the "Tartars"; a progress to Greenwich, where messages of approval from Elizabeth are received; the presentation of Francis Davison's *Masque of Proteus* at court; as well as banquets, masques, theatrical devices, dinners, fights at barriers, seemingly endless occasions for general reveling, and a great deal of music and dancing.[9] A few passages from the opening pages of the text will serve to provide an indication of the flavor of the events:

> The great number of gallant Gentlemen that *Grays-Inn* afforded at ordinary Revels, betwixt *All-hollontide* and *Christmas*, exceeding therein the rest of the Houses of Court, gave occasion to some Well-willers of our Sports, and Favourers of our Credit, to wish an Head answerable to so noble a Body, and a Leader to so gallant a Company: Which Motion was more willingly hearkened unto, in regard that such Pass-times had been intermitted by the space of three or four Years, by reason of Sickness and Discontin-uances. [*GG*, 5]

In response, and after some deliberations among themselves, a Lord of Purpoole is named:

> Whereupon, presently they made choice of one Mr *Henry Helmes*, a *Norfolk*-Gentle-man, who was thought to be accomplished with all good Parts, fit for so great a Dignity; and was also a very proper Man of Personage, and very active in Dancing and Revelling. [*GG*, 6]

and the "King at Arms" formally presents the Prince:

> I do pronounce my Sovereign Liege Lord, Sir Henry, rightfully to be the high and mighty Prince of Purpoole, Arch-Duke of Stapulia and Bernardia, Duke of the High and Nether Holborn, Marquis of St Giles's and Tottenham, Count Palatine of Bloomsbury and Clerkenwell, Great Lord of the Cantons of Islington, &c. Knight of the most honourable Order of the Helmet, and Sovereign of the same. [*GG*, 14]

While there may be more than a hint of undergraduate exuberance here, there is nevertheless a certain seriousness of purpose. For Bland, this seriousness is effec-tively an extension of the training mission of the Inns, broadly conceived:

> *Gesta Grayorum* was no ordinary undergraduate rag. Though this element is not absent, the revels were also intended as a training-ground in "all the manners that are learned by

---

[9] Bland offers an inventory of the events of the Revels (*GG*, xiv–xvi). A more detailed inventory, together with a sustained analysis, is on offer in Knapp and Kobialka (1984).

the nobility". In the words of a mid-sixteenth century report on the Inns, "this is done only that they should come to know how to use themselves". Dancing, music, declamation, acting, the etiquette of a formal procession, on foot or on horseback, the proper ordering of a banquet, the exchange of courtesies in speech or in writing, all these are essential parts of the elaborate make-believe. And they are there because this kind of behaviour was expected of the public figures of the age, among whom, in due course, these young men would take their place. [*GG*, xxiv–xxv][10]

And in any event, one can gauge this seriousness by considering the attention the Revels garnered from queen and from court. "On the 3d. of January at Night" (the evening of the staging of Bacon's *Device*), the *Gesta Grayorum* reads, "there was a most honourable Presence of Great and Noble Personages, that came as invited to our Prince." It continues, with elaborate specificity:

> the Right Honourable the Lord Keeper, the Earls of Shrewsbury, Cumberland, North-umberland, Southampton, and Essex, the Lords Buckhurst, Windsor, Mountjoy, Sheffield, Compton, Rich, Burleygh, Mounteagle, and the Lond Thomas Howard; Sir Thomas Henneage, Sir Robert Cecill; with a great number of Knights, Ladies and very worshipful Personages: All which had convenient Places, and very good Entertainment, to their good Liking and Contentment. [*GG*, 35]

What the *Gesta Grayorum* preserves for us, then, is a highly detailed account of all manner of entertainment ("Sports," to use one of the favorite terms from the text itself) that was understood—at least in some of its parts or some of its dimensions—to be something like serious play. As such, the text stands as a remarkable document that bears the marks of a range of important early modern practices: reveling and dance, to be sure, together with archly self-conscious theatrical play (both on and off literal stages), but also at least the broad outlines of a certain awareness of class and the expectation of class-based opportunities in the near future for which not only legal training but also performance and mimicry were understood to be integral preparation.

The Revels and its textual embodiment in *Gesta Grayorum*, in all the great variety of its components—set speeches, controversies, progresses, theatrical performances, masques, fights at barriers, and visits to the court of Elizabeth—has been of interest to literary historians and literary critics and their work has constituted, in general terms, the attempt to reach a greater understanding of the Revels within the particular context of early modern literary culture.[11] W. W. Greg's 1915 edition of the text produced for the Malone Society, for instance, offers the reader an explicit identification of the literary importance of the *Gesta Grayorum*: "There are three main points of literary interest in the *Gesta Grayorum*, namely, a supposed allusion to Shakespeare's *A Comedy of Errors*, the speeches of the six Councillors, and the *Masque of Proteus*." For each of these points, Greg offers a brief orientation

---

[10] Bland's reference to a sixteenth century report is to Waterhouse (1663), a work that contains what Bland describes as a "report on the Inns presented to Henry VIII by Nicholas Bacon and others in about 1550" (*GG*, xxv).

[11] See West (2008), Elliott (1992), O'Callaghan (2007), Magnusson (2004), Corrigan (2004), and Lanier (1993).

and summary: the allusion to *The Comedy of Errors* is (for Greg at least) of dubious validity; the "interest of the Councillors" speeches and the Prince's reply . . . lies in the fact that they have been attributed to the pen of Francis Bacon'; and "the portion of the work which is really the occasion of the present reprint" (Greg, vi), the *Masque of Proteus*, for which the poet Francis Davison is "the main author" (Greg, viii), stands as the literary centerpiece of the entire work.[12]

In order to illustrate briefly the richness of these discussions of the literary significance of the *Gesta Grayorum* in the years since Greg, I will point to two particularly significant instances: Stephen Orgel's discussion of *The Mask Of Proteus* from the Revels and Bruce Smith's discussion of what he calls "male enterprise" in the coincidence of Shakespeare's *A Comedy of Errors* and its first recorded performance in the Gray's Inn Revels on 28 December 1594.[13] I hope to tell through this brief exercise two important—and related—stories: the first is the story of the literary culture of early modern England that can be said to give rise to the Revels. The second story tells of the emergent culture of early modern science that, while distinct in its specificity, is nevertheless deeply implicated in the literary culture of the period. As will become clear in the following pages, what is crucially at stake in this discussion is the recognition of the mutually informing and sustaining work undertaken in both analyses of the literary and the scientific densities of the events organized and celebrated in the *Gesta Grayorum*. In other words, my point is not to separate Orgel's discussion of the *Masque of Proteus*, for instance, from what I would like to identify as the coordinated status of the *Masque* within the culture of science. In this regard it becomes clear exactly how socially and culturally embedded we find the literary and scientific cultures of the period.

Francis Davison's *Masque of Proteus and the Adamantine Rock* was performed at the Banqueting House at Whitehall on 3 or 4 March 1595, where (as described in Orgel's account), upon entering, the queen and courtiers saw "a simple scaffold stage at one end, with the adamantine rock of the title as a prominent fixture" (p. 9).[14] The masque opens with a hymn in "prayse of *Neptunes* Empery" (*GG*, 77) sung by sea nymphs and includes four speaking roles: the Prince of Purpoole's Squire, Proteus, Amphitrite, and Thamesis.[15] The Squire begins by announcing his surprise that Proteus should have kept his promise ("I rather deemd & not without

---

[12] Bland agrees with this assessment: "From a strictly literary point of view it is the *Masque of Proteus* which is the most interesting and important feature of the *Gesta Grayorum*" (*GG*, xxiii).

[13] Orgel (1965: esp. 8–18); Smith (1997: 102–25).

[14] Orgel continues: "The stage had no curtain and no proscenium arch, and there is no reason to assume that the rock, from which the masquers issued forth, was anything more elaborate than a painted canvas flat large enough to conceal eight people" (p. 9). For a detailed discussion of the hall as playing space, see Knapp and Kobialka (1984).

[15] The opening hymn ("Of Neptunes Empyre let us sing") was reprinted by Davison in his anthology of poems entitled *Poetical Rhapsody* (1602) and is attributed there to Thomas Campion. For a discussion of the textual state of *The Masque of Proteus* as printed in the *Gesta Grayroum* and its relation to the manuscript version in the British Library (Harley 541), see Bland (1968: xi). Greg prints both versions in his edition and adds two interesting details: the Harley manuscript "is said to have formed part of the collections of John Stow, the antiquary, and the text of the masque is a caligraphical exercise of considerable beauty" (Greg [1914: vii]). Greg reproduces a facsimile of a page from Harley 541 (fol. 139a).

good cause,/That those still floating regions where you bide,/And th'ever-changing nature that you have./ Nought els but breach of promise promised" [*GG*, 78]). In the ensuing dialogue, and in response to Amphitrite's request for an explanation of the particular promises under discussion, the Squire tells the story of events that lead to the present moment and Proteus's appearance, together with the adamantine rock. Upon his return journey from helping the king of Russia defeat his "Tartarian" enemies (*GG*, 79), the Prince of Purpoole, "one Sunshine day/Resting him self within a goodly tuft,/Of tall straite-fir-trees that adornede that shore,/ Reading a letter, lately sent unto him" (*GG*, 79), notices a "great unusuall flock" of porpoises; he draws near the shoreline to see better and stumbles upon a cave "whose frame seemd more then naturall."[16] The Prince enters and spies the sleeping Proteus and, "being neer with great agility/Seasd suddenly upon this Demy God." True to form (one could say), Proteus "resorts presently/To his familiar artes and turning tricks" in order to regain his liberty. He transforms first into "a goodly lady passing faire" (*GG*, 80), then into a serpent, then into a great treasure chest of "many Diamonds/& Rubies of inestimable worth," then lastly (and most disturbingly) into the "sad spectacle" of one of the Prince's favorite knights "Mangled and pierst with many a grisly wound" (*GG*, 81). None of these transformations, however, defeats the Prince, who refuses to release Proteus from his grasp. Proteus responds by offering ransom: the Prince's fortune foretold, "huge treasures, Ladies loves,/Honour, and fame of famous victories" (*GG*, 81); all are rejected. Proteus's final offer is his promise to remove his Adamantine rock ("The seas true star") from under its hiding place "under Th'artik pole" and relocate it wherever the prince should determine apt. Proteus assures the Prince

> That the wide Empire of the Ocean,
> (If his fore telling spirit faild him not,)
> Should follow that, wheare ere it should be sett. [*GG*, 82]

But there is one proviso: "That first the Prince should bring him to a power,/Which in attractive virtue should surpas/The wondrous force of his Ir'ne drawing rock" (*GG*, 82). The Prince (the Squire informs us) is supremely confident he can indeed win the challenge and offers himself and seven of his knights as collateral. The wager accepted, the Prince and his knights enter into the Adamantine rock as Proteus's hostages until "this great Covenant should bee performd,/Which now rests to bee done" (*GG*, 82). With this, we return to the present moment of the masque and the Squire calls upon Proteus to "Blazon forth" the virtues of the Adamantine rock.

---

[16] The letter the Prince reads was sent to him by one of his knights and tells of the gift he is sending to the Prince: "a Commoditie/Of Pigmeys taken in his private quest" (*GG*, 80). These "Pigmeys" make an appearance later in the masque, leading the principal figures to the dance that stands as the culmination of the action. They also bear the masquers heraldic emblems: "the Pigmies brought eight Escutcheons, with the Maskers Devices thereupon, and delivered them to the Esquire, who offered them to Her Majesty" (*GG*, 86).

Proteus's blazon is an interesting testament to a certain understanding of the
supremacy of iron, "prince by auncient right"—with power to command "Rebel-
lious golde" (*GG*, 83):

> Thus hath this Load-stone by his powerful touch
> Made th'Iron Load-star of the world,
> A Mercury to point the gainest way
> In watry wildernes and the desert sands.
> In confidence whereof th'assured Mariner
> Doth not importune Jove for sin or stars
> By this Attractive force was drawne to light
> From depth of ignorance that new-found world
> Whose golden mines Iron found and conquered. [*GG*, 83]

The Squire, however, is unmoved by Proteus's blazon of the virtues of the adamantine rock and its power to "forceth yron that all things els comands" (*GG*, 82).
Instead, the Squire advises Proteus to "calme awhile your overweening vaunts/And
prepare beleefe & doe but use your eyes." Then begins the Squire's own blazon—in
this case, in praise of Elizabeth, "Excellent Queene, trew adamant of Hartes":

> What can your Iron doo without Armes of men,
> And armes of men from hartes of men doo move,
> The hartes of men, that's it thence motion springs
> Lo Proteus then Th'attractive Rock of hartes,
> Hartes which once truly touced with her beames
> Inspiring purest zeale and reverence
> As well unto the person as the Power,
> Do strayt putt of all temper that is false,
> All hollow feare and skooled flattery
> Turne fortunes wheele, they ever keepe their course,
> And stand direct upon the Loyal line. [*GG*, 84]

The Squire continues in this vein, enumerating the many ways in which Elizabeth
is more truly attractive than Proteus's adamantine rock—not least in the ways in
which she has inspired the "Princes high/Of forreigne lands" who have "vowed
pilgrimage"; or in her own emergent empire, "As Russia, China, & Magellanus
straytes/Can witnes bear." In the very face of this vision of Elizabeth, Proteus is
converted: "Blest be the Prince that fors't mee see this grace,/Which worldly
Monarkes & Sea-powers adore" (*GG*, 84–5). He then strikes the rock with his
"bident," releasing the Prince of Purpoole and his seven knights "in a very stately
Mask, very richly attired, and gallantly provided of all things meet for the performance of so great an Enterprize" (*GG*, 86).[17]

---

[17] The praise of the monarch to which *The Masque of Proteus* is generically dedicated seems to
migrate—from the masque itself to the concluding paragraph of the *Gesta Grayorum* in which the great
triumphs of the Revels are noted, but only to emphasize how utterly they pale in comparison to Queen
Elizabeth:

> But now our Principality is determined; which, although it shined very bright in ours, and other
> Darkness; yet, at the Royal Presence of Her Majesty, it appeared as an obscured Shadow: In this,
> not unlike unto the Morning-star, which looketh very chearfully in the World, so long as the Sun

In Orgel's important book on the Jacobean masque, Davison's *Proteus* is significant in a number of ways: it is the last Elizabethan masque for which we have a complete text; it can be said to embody those defining elements of the court masquing tradition inherited by the Stuart creators (especially Jonson); and it is the first of the Tudor entertainments that "at all resembles the standard Jacobean masque" (Orgel [1965: 8]). Among its forward-looking traits, one is particularly relevant: its non-traditional staging. "*The Mask of Proteus*, with its fixed stage and its unified setting, is the first English masque to conceive, in however small a way, of the masquing hall as a theater" (p. 9).[18] The understanding of the essentially *theatrical* (rather than dramatic) nature of the contest staged between Proteus and his rock, and the Squire's not just invocation but *presentation* of Elizabeth herself there in the hall, is not only crucial to the masque, but is also its very definition.[19] The masque form depends upon its hero—and the masquers and the spectators—knowing that there is a world *outside* the fiction. After the elaborate blazon of the rock, the Squire easily defeats Proteus by directing "Proteus's" attention outward, beyond the limits of the stage and the fictive world, toward their royal spectator. He invokes Elizabeth, more powerful than the lodestone, "true Adamant of hearts" (p. 13). It is this move beyond the stage to the larger stage of the court that allows for Proteus's defeat and conversion. The Squire's victory is enabled by a theatrical self-consciousness. Proteus, with his brand of literalism, stands "helpless before a figure who can step outside that world and who can see the properties of the rock as metaphorical rather than physical—a figure, in short, who knows that he is an actor in a masque and is conscious of the presence and significance of the audience" (p. 13). On this account, the masque completely depends upon the relationship between the literal and the metaphorical: Proteus's literal faith in the lodestone, on the one hand, and the Squire's faith in the metaphorical power of the Queen's attraction. Indeed, this tension abides within the structure of the masque, in the "artistic self-consciousness whereby the virtue or power of a figure is dependent on his ability not to take himself or his world literally" (p. 14).[20] But within the world that the masque imagines, this relationship between the literal and the metaphorical is more complicated still:

looketh not on it: Or, as the great Rivers, that triumph in the Multitude of their Waters, until they come unto the Sea. *Sic vinci, sic mori pulchrum.* [*GG*, 88–9]

[18] This is by no means to deny the fundamentally *theatrical* nature of Elizabethan entertainments and pageants—or, even, the coronation entry, for that matter—all of which certainly deployed theatrical effects quite deliberately and brilliantly and which succeeded in recasting the very streets of London into theatricalized spaces. It is, however, to draw attention to the gesture toward the theater as purpose-built space (the theater as theater) that (as Orgel argued) is established in Davison's masque.

[19] The masque is unlike the play—its dialogue, for instance, is intended "to *delay* the resolution," whereas in the drama, dialogue works toward resolution. In the masque, resolution is provided only by dance and music: "Indeed . . . it is characteristic of the kind of action masques present that it can take place only in a world purged of drama, of conflict" (Orgel [1965: 16–17]).

[20] This is a power, Orgel argues, that is rare "outside the masque." But there are "analogues" elsewhere in early modern literature: Shakespeare's Prospero, for example, Spenser's Sir Guyon (*Faerie Queene*, IV.i.3), and Middleton's De Flores (who plans for a murder *between acts*). "Hero or villain, the figure who is capable of stepping outside or violating the conventions of his form is by nature invincible; and that figure in Davison's masque is the Esquire" (Orgel [1965: 15]).

In part, of course, Proteus loses in the trial through a lawyer's trick, a play on words. But in the world of the masque, the trope also expresses a literal truth; the metaphorical "Adamant of hearts," the queen's "attractive virtue," has the same power as the lodestone. [p. 15]

The contest between the figural and the literal is also at stake in Bruce Smith's discussion of the *Gesta Grayorum* and *The Comedy of Errors*—two related works that "from the standpoint of cultural anthropology" can be said to "constitute a single datum, a single piece of social process" (Smith 1997: 108). Smith's point of departure is the status of the Gray's Inn Revels as an elaborate rite of passage undertaken by its specifically male participants. As such, the *Gesta Grayorum* (as the chronicle of this rite) also documents (as it were) the three phases identified by Arnold Van Gennep as essential to such rites—separation, transition, and incorporation—phases in which the individual re-negotiates his identity by first divesting himself of a former identity, then by accepting a new identity, and finally by affirming that new social identity in a specifically public way.[21] Smith's commentary on the third phase—incorporation—is especially significant to the present discussion, in part because incorporation in the *Gesta Grayorum*, Smith argues, is marked by the *Masque of Proteus*, even as incorporation (the act through which "the subject is, literally, 'embodied' in the society of which he has now become a member" [p. 116]) is marked in *The Comedy of Errors* in the play's final scene of recognitions and reconciliations. And if in the elaborate conclusion to *The Comedy of Errors* "artifice and illusion have been deployed in service to a higher truth," the case is even more pronounced in *Proteus*. For Smith, this mutual deployment in Davison's masque is a matter not merely of the relationship between artifice and reality, but rather the generic nature of the masque itself as it comes to embody the transformation from the one to the other:

As an elaborate invitation to the dance, a mask finds its true end when the art quite literally turns into reality and the maskers go out into the hall, choose partners from among the spectators and invite those partners to dance . . . In this ritualized act of homage, the make-believe court of the Prince of Purpoole not only defers to the real-world court of Queen Elizabeth, it dissolves into it. [pp. 117–18]

For Smith, the similar ways in which *The Comedy of Errors* and the *Gesta Grayorum* end serves "to reify the artifice, to reveal that the artifice is, after all, reality itself. Play has all along been doing social *work*" (p. 118).[22]

---

[21] See van Gennep (1950). Smith also references V. Turner (1982).

[22] One particularly important manifestation of the social work that Smith sees in process, as it were, in Shakespeare's play and the Gray's Inn Revels is the subject of Rivlin (2002). Rivlin is especially interested in tracing the emergence of competing versions of literacy—traditional formal literacy, on the one hand, and a new functional literacy, on the other—and the consequences the virtually inevitable clash between them had in the period. On Rivlin's account, the theater "provided a space in which questions of the correspondence between external sign and internal reality could be explored." Theatrical literacy, then, emerges as "the confluence of a formal literacy still accessible and applicable mostly to privileged males with a broader notion that includes cultural competence and instrumentality" (p. 65). Rivlin discusses these dynamics within both Shakespeare's play and the Gray's Inn Revels, particularly the "Night of Errors." For Rivlin, these two texts together:

For the moment I would like simply to mark the argument (which I will address more directly below) that is shared by both of these literary (or literary critical) arguments. For both Orgel and Smith (whose work here is offered here as illustrative, as particularly powerful examples of their general kind), the success of Davison's *Proteus*—and, to a significant degree Shakespeare's *A Comedy of Errors* and, even, the Gray's Inn Revels on the whole—depends upon the active renegotiation of the conventional understanding of the relationship between artifice and nature. And it is this renegotiation that is also shared by readers, critics, and historians across a range of disciplines. Indeed, even as the *Gesta Grayorum* has been important to literary historians and critics (in ways I have sketched quickly above), so has it been of interest to critics and historians interested in scientific culture in the early modern period. We can perhaps see a hint of this trajectory in Orgel's discussion of *Proteus*. Regarding the Squire's notion of the potential deafness of sea creatures (in the opening Hymn), Orgel writes:

> The legal mind at play works in amusingly literalistic ways. The Esquire's response to the opening song is to wonder at its being produced by minions of Proteus; for philosophers hold that sea creatures have no interest in music and, indeed, may very possibly be deaf. To this Aristotelian hypothesis Proteus returns a curt Baconian reply, suggesting that philosophers try living under water before they theorize about it and citing as a piece of empirical evidence the experience of the dolphin charmed by Arion's music. [pp. 11–12]

But where critics concerned with science and the *Gesta Grayorum* locate their discussions differs: while literary critics tend to focus on either the Shakespearean discipline (as identified above) or the complications surrounding *The Masque of Proteus*, critics interested in early modern science understandably tend to focus attention on that portion of the *Gesta Grayorum* understood to have been written by Bacon—and therefore to represent Bacon's efforts to reinvent natural philosophy. In the *Device*, the king announces his intention to seek from his six favorite counselors advice:

> not [on] any particular Action of Our State, but in general, of the Scope and End whereunto you think it most for our Honour, and the Happiness of Our State, that Our Government be rightly bent and directed ... And this We require you to do, without either Respect to Our Affections, or you own; neither guessing what is most agreeable with Our Disposition, wherein We may easily deceive you; for Princes Hearts are inscrutable [*GG*, 44].[23]

The first counselor advises the exercise of war and "Conquest." He asks the king: "What is your Strength, if you find it not? Your Fortune, if you try it not? Your

---

challenge the idea that literacy grounds character, recognizing literacy instead as both an unstable signifier and a culturally uncontrolled (and perhaps uncontrollable) tool. Both show the ways in which participation in literate culture might be diffused and made instrumental in shaping character. By obfuscating the direct correspondence between formal literacy and social rank, the play and the record of its early performance render inscription non-transparent and trouble easy taxonomic schemes. [pp. 74–5]

[23] For a brief discussion of the attribution of the *Device* to Bacon, established by Spedding in his edition of Bacon's works, see Vickers's discussion in Bacon (1996: 532).

Vertue, if you shew it not?" And as for the outcomes of conquest, the first counselor promises the king he will insure his own lasting reputation, leaving behind "deep Foot-steps of your Power in the World" (*GG*, 45–6).

The third counselor (for I will come to the second one last, since it is his advice that we recognize as Bacon's own) argues the king should strive for "Eternizement and Fame, by Buildings and Foundations." This counselor rehearses what he sees as the fundamental and disqualifying "Difficulties and Errors in the Conclusions of Nature" and the fall from "these immoderate Hopes and Promises" of "Mystical Philosophy into Comedies of ridiculous Frustrations and Disappointments of such Conceipts and Curiosities" (*GG*, 48). For his part the fourth counselor argues for the pursuit of the "Absoluteness of State and Treasure," since (as he sees it) "nothing causeth such dissipation of Treasure as Wars, Curiosities and Buildings" (*GG*, 50–1).

The fifth counselor argues for the pursuit of "Vertue, and a Gracious Government" (*GG*, 52) and offers a vision of what seems to be a perfectly reasonable call to virtuous—if patriarchal—government based on justice and fairness. The sixth and final counselor rejects the previous five and their advice—some for thinking of the Prince "as of a Prince past" (*GG*, 54) and others (four and five) for their "cumbersome" lessons, "as if they would make you a King in a Play; who when one would think he standeth in great Majesty and Felicity, he is troubled to say his part" (*GG*, 55). This counselor then makes clear his own preferences for "Pass-times and Sports" (*GG*, 54): "What! Nothing but Tasks, nothing but Working-days? No Feasting, no Musick, no Dancing, no Triumphs, no Comedies, no Love, no Ladies? Let other Men's Lives be as Pilgrimages, because they are tied to divers Necessities and Duties; but Princes Lives are as Progresses, dedicated only to Variety and Solace" (*GG*, 55). He continues:

> And therefore leave your Wars to your Lieutenants, and your Works and Buildings to your Surveyors, and your Books to your Universities, and your State-matters to your Councellors, and attend you that in Person, which you cannot execute by Deputy: Use the Advantage of your Youth ... make your Pleasure the Distinction of your Honours ... And, in a word, *Sweet Sovereign*, dismiss your five Councellors, and only take Councel of your five Senses. [*GG*, 55]

The sixth counselor makes clear that he understands the others counselors to speak a certain kind of nonsense. His recommendation, by contrast, suggests the Prince literally embrace the world of the senses through the pleasures they can provide. Among the other counselors whose advice is thus so easily dismissed in favor of pleasure, it is only the second one who has addressed the senses in any way. But he does so neither to celebrate them nor to repudiate them (one can easily imagine a spiritual counselor to the Prince who may well have made such a case), but rather to discipline the senses to the work of the mind. And, indeed, it is this counselor who argues for "the Study of Philosophy" that functions effectively as Bacon's spokesperson. He "wish[es] unto your Highness the Exercise of the best and purest part of the Mind, and the most innocent and meriting Request, being the Conquest of the

Works of Nature" (*GG*, 46). By this path, the counselor continues, and as a result of "bend[ing] the Excellency of your Spirits to the searching out, inventing and discovering of all whatsoever is hid in secret in the World," the prince will "be not as a Lamp that shineth to others, and yet seeth not it self; but as the Eye of the World, that both carrieth and useth Light" (*GG*, 46–7).

The "philosophy" that this counselor recommends will become well known in Bacon's lifetime—and, in some measure, perhaps due to Bacon's leading efforts in giving programmatic focus to a generalized desire at play across the culture—as the study of nature and, more particularly, *causality* in nature. In light of the historical precedents for a state's pursuit of philosophy—Persians, Greeks, Egyptians ("those Kingdoms were accounted most happy, that had Rulers most addicted to Philosophy" [*GG*, 47]), the second counselor "commend[s] to your Highness four principal Works and Monuments of your self." First: "The collecting of a most perfect and general Library." Second: "a spacious, wonderful Garden" which will include rooms and rare beasts, a fresh water and salt water lake for various fishes, etc., so "you may have, in a small Compass, a Model of Universal Nature made private" (*GG*, 47). Third: "A goodly huge Cabinet, wherein whatsoever the Hand of Man, by exquisite Art or Engine, hath made rare in Stuff, Form, or Motion, whatsoever Singularity, Chance and the Shuffle of things hath produced, whatsoever Nature hath wrought in thing that want Life, and may be kept, shall be sorted and included" (*GG*, 47–8). And fourth: "Such a Still-house so furnished with Mills, Instruments, Furnaces and Vessels, as may be a Palace fit for a Philosopher's Stone" (*GG*, 48). The outcome that this counselor predicts for such a king as would pursue this new agenda for the apprehension of natural knowledge is indeed an exulted one:

> Thus when your Excellency shall have added depth of Knowledge to the fineness of Spirits, and greatness of your Power, then indeed shall you lay a *Trismegistus*; and then when all other Miracles and Wonders shall cease, by reason that you shall have discovered their natural Causes, your self shall be left the only Miracle and Wonder of the World. [*GG*, 48]

Bacon's dream of a philosopher-prince who would literally embody the processes of natural philosophy's illumination of nature's secrets and nature's causes was an enduring one; it is visible in Bacon's earliest writings (in another entertainment, *Of Tribute*, for example), and leads to Bacon's fantasy in his *New Atlantis* of an entire—and ideal—state literally constructed on the model of the sovereignty of a natural philosophy expressed in the very structure and disposition of the state itself.[24] And although this dream—which reaches its highest articulation in the mid-seventeenth-century founding of the Royal Society—would ultimately prove to be nothing more than a dream, Bacon's fundamental vision of the promise held out by the systematic and rigorous pursuit of nature's causes would prove generative outside of the court and beyond the prince, in the birth and growth of multiple

---

[24] *Of Tribute* and *New Atlantis* are both reprinted in Bacon (1996: 22–51 and 457–89 respectively).

sites and practices devoted to the development of natural philosophy.[25] In fact, the four "action items" (as it were) proposed by the second counselor—the establishment of a library, a garden, a "cabinet," and a "still-house"—are precisely indicative of this "private" development of early modern science. It is, moreover, indicative of four important localizations—or sites of emergence, as I would like to call them— of particular domains or fields of scientific activity and pursuit.

But at the same time that Bacon's *Device* advocates these particular localizations or sites for the development of what will emerge as science, it is also *itself* a cultural performance that arises only within a particular social, political, and economic setting. The text of the *Gesta Grayorum* bears all the marks of this kind of local and contingent emergence. Arranged across the Christmas holidays, 1594–5, the revels bore a particular relation to the calendar and were cyclical in nature (in that they were enacted annually, though not without the occasional hiatus). They were also organized and structured in a particular relation to political and cultural institutions (the monarchy, most significantly) and practices (entertainments, theatricals, dances, etc.) and were akin to the pre-Lenten traditions of carnival, including the ritual election of a Lord of Misrule, in this case, Henry, Prince of Purpoole.[26]

As is perhaps not surprising, these speeches—and, of course, the second counselor's in particular—have drawn the attention of numerous critics interested in early modern science, whether that interest takes the form of a focus on Bacon in particular, or is part of a more general concern with a growing desire in the period to rethink the nature of the empirical study of nature and the production of natural knowledge. Thus, for instance, Paolo Rossi sees Bacon's *Device* as perfectly consonant with his sustained efforts first to distance himself from "arbitrary uncontrolled personal research," and then to turn instead to "organised collaborative experiment."[27] For Rossi, Bacon "pursued his plan of scientific reform with astonishing perseverance and very little success" (p. 23). As evidence of this perseverance, Rossi points first to Bacon's frequently cited 1591 letter to Burghley in which he

---

[25] The birth and development of (non-governmental) institutions, societies, and even less formal associations and gatherings dedicated to the study of nature and causality in nature was a continent-wide phenomenon. For an extended discussion of one of these—the Academy of Linceans, founded by Federico Cesi (and which included Galileo among its members)—see Freedberg (2002).

[26] By its very title the *Gesta Grayorum* makes reference to an earlier collection of moral tales, the *Gesta Romanorum*. Derived from tales and stories in Latin and assembled in the late thirteenth or early fourteenth century, the allegorical and moralizing tales of the *Gesta Romanorum* were enormously popular and influential. The various compilations of these stories and their textual embodiment in printed versions was altogether a fluid matter, but one particular textualization, assembled and translated by Richard Robinson, was published in London in 1595 (and again in 1602 and 1610) and was widely popular. Richardson's text carried the title *A Record of Auncient Histories, intituled in Latin: Gesta Romanorum*; its extended title included the phrase "Discoursing upon sundry examples for the advauncement of vertue, and the abandoning of vice" and in it perhaps we hear a formulation that Bacon himself later would adopt into the title of one of his great texts, *The Advancement of Learning*. In any event, it is clear that these texts capture not only in title, but also in spirit, two contrary trends: the movement, on the one hand, toward the reiteration and repetition of allegorical tales for moral training, and, on the other, the critical shift away from inherited wisdom (even if it comes in the form of short narratives) toward a forward-facing interest in the new and the novel.

[27] Rossi (1978: 23).

famously confessed that he had "as vast contemplative ends, as I have moderate civil ends" and then announces his great ambition:

> for I have chosen all knowledge to be my province; and if I could purge it of two sorts of rovers, whereof the one with frivolous disputations, confutations, and verbosities, the other with blind experiments and auricular traditions and impostures, hath committed so many spoils, I hope I should bring in industrious observations, grounded conclusions, and profitable inventions and discoveries; the best state of that province. [p. 20]

Rossi offers further evidence in the form of the *Gesta Grayorum* (which, unaccountably, he calls Bacon's "speech entitled *Gesta Grayorum*"), with special emphasis on the proposed institutional nature of Bacon's reformed science. In the *Device*, Rossi observes, Bacon "gave a detailed description of his programme of reform: it was not concerned with individual discoveries and inventions but it required scientific establishments and institutions" (p. 23). The fact that Bacon ultimately was unsuccessful is not the point, for Bacon's "plan for the reform of science was his great contribution to culture" (p. 25). And this effort was indeed a sustained one across Bacon's career—from these early pronouncements through *The Advancement of Learning* and to, in the end, his scientific utopian fantasy the *New Atlantis*.

Bacon's relatively early efforts on display in the *Device* also figure importantly—but differently—in Paula Findlen's essay "Anatomy theaters, botanical gardens, and natural history collections," which begins with Bacon's *Device*, identifying it as one instance in which Bacon "began to fantasize about the locations for knowledge."[28] While Rossi is interested in the institutional (rather than the individual) nature of the science Bacon sought to establish, Findlen takes a closer look at the kind of work that was undertaken by individuals within particular sites dedicated to natural philosophy. Citing the second counselor's description of the library, the garden, the cabinet, and the laboratory, Findlen provides this useful summary:

> This statement suggested a new idea of empiricism that privileged human invention and demonstration over pure observation and celebrated the communal aspects of observing nature over the heroic efforts of the lone observer. Nature had to be reconstructed within a microcosm, creating an artificial world of knowledge in which scholars prodded, dissected, and experimented with nature in order to know it better. [p. 272]

Findlen also points to Bacon's "final" iteration of this fantasy in the *New Atlantis*, and in the vision of Salomon's House, in particular, that "knowledge-making center of the realm":

> Surrounded by artificial mines, lakes, a botanical garden, and a menagerie . . . it represented a full elaboration of science as an activity that removed nature *from* nature in order to study it better. Bacon's remarkable array of unique spaces for science mirrored the variety of possible experiences that one could have of nature, isolating all natural objects and processes. [p. 273].

---

[28] Findlen (2006: 272–89).

Although Bacon's various fantasies of such scientific institutions remained fantasies, they were nevertheless not only forward-looking (as Rossi had suggested), but also well-grounded in recent and contemporary innovations in science. Findlen points to the real-world construction of "purpose-built spaces" in which the new work of science was undertaken, arguing that they all "acted in ways similar to Bacon's utopian vision of science; to differing degrees, they removed natural artifacts from their original locations, placing them inside new spaces for the specific purpose of studying them in order to improve natural knowledge" (p. 273). It was this move to render the natural artificial that constitutes Bacon's great contribution: "If understanding nature was a kind of art," Findlen writes, "then all of the ways in which nature could be rendered artificial became important in developing a complete understanding the natural world" (p. 289).

With this assertion, we return, as it were, to the argument marked above in consideration of the literary critical works by Orgel and Smith on aspects of the *Gesta Grayorum*. Even as the success of the literary work staged in the Revels depends upon the renegotiation of the relationship between the natural and the artificial, so too does that fundamental renegotiation inform the work of historians and critics of early modern science. As Findlen's notion of "the ways in which nature could be rendered artificial" suggests, one of the most important of the challenges we face in our efforts to understand early modern science requires that we establish precisely this analytical imperative to rethink the nature–artifice relationship.

In her recent study *The Jewel House: Elizabethan London and the Scientific Revolution*, Deborah E. Harkness, too, focuses on the issue of the various sites and locations in the period that were devoted to various pursuits of natural philosophy. Harkness offers a vision of Bacon's fantasies of reform and the development of institutions dedicated to the production of natural knowledge that existed in a tense relationship with efforts toward these goals already underway in the city of London, though these efforts are (for a complex of historical reasons) less commonly known today. On Harkness's account, Bacon's desire to institutionalize science was, in part, a reaction against what she describes as "London's doubtlessly energetic—but to his mind inchoate and purposeless—inquiries into nature." And Bacon's task, Harkness writes, was to find a way to transform these inquiries "into a tool of state that could benefit the commonwealth."[29]

Harkness argues that, with the exception of his having located all of the scientific pursuits of natural knowledge "within a single, hierarchical institution overseen by a single, well-educated man," all such activities as Bacon imagines in Salomon's House "already existed in the City of London." Our contemporary notion of Bacon's strictly visionary nature, then, as it relates to the reformation of science along institutional lines, is a trick of historical perspective:

> It was not until the end of the seventeenth century, when the memories of the Elizabethan interest in nature had faded and the Royal Society had been established,

---

[29] Harkness (2007: 7).

that people began to look back on Bacon as a prophet of a newly empirical science. And their view has shaped our view. The intervening centuries have not been kind to Elizabethan London's interest in the natural world, and our knowledge of it has been so slight that we, too, saw Salomon's House as a blueprint for what science could *become* rather than a description of what science already *was*. [p. 7]

In the final chapter of her book, Harkness elaborates her critique of Bacon by contrasting the programmatic work he (merely) proposed—initially in such works as the *Device* from the *Gesta Grayorum* and then, in revised forms, in *The Advancement of Learning* and the *New Atlantis*—to the scientific work actually undertaken and to a certain extent realized by Hugh Plat, as represented in his notebooks and in his book, *The Jewel House of Art and Nature* (1594). For Harkness, Bacon's program for the reformation of science is fundamentally a matter of "power" (whereas Plat's program was largely a matter of "process" [p. 242]) and this fundamental concern is on conspicuous display in the Gray's Inn *Device*. This "masque" (as Harkness curiously—and unclearly—names it), "sounded preachy" and moreover was not, she assures us, "enthusiastically received by his audience" (p. 213).[30] But it was not forgotten, for Bacon returned to these "ideas about institutions for the study of nature" in his *New Atlantis*, particularly in Salomon's House. In this later version, "the *Gesta Grayorum*'s small clutch of scientific spaces—library, garden, cabinet, and laboratory—were expanded into a complex as sprawling as Bacon's Inns of Court" (p. 213). Harkness inventories a number of these spaces: specially equipped caves, hospitals, distillation apparatuses, orchards, gardens, sound houses, perfume houses, bake houses, brew houses, engine houses, among others. But Salomon's House, Harkness concludes, "was not a wishful romance. Instead, it was a dressed-up representation of the real world of science in Elizabethan London" (p. 213).

In many ways, Harkness's argument against Bacon—or, more precisely, her argument against his unearned status and reputation as the putative founder of modern science—is an argument against the very idea of such easily identified origins of something as vast, heterogeneous, and complex—in short, for something that is *cultural* in nature—as is early modern science. Harkness is responding, of course, to a more general feature of conventional histories of science, particularly of the English tradition within which the Renaissance is often figured as the pivotal moment. These histories offer evidence of the long-standing desire to discover the origin of modern science, that more or less generally identifiable point in time after which science is imagined to be so thoroughly autonomous of nature (its normative object of investigation) and culture (its proper background) that both effectively fall away, or at least into clearly subordinated positions that do not trouble what is sketched as the inevitable progress of science across time. Such identified origins

---

[30] The identification of the *Device* as a masque is interesting, especially set against Bacon's later dismissive account of such entertainments in his essay, "Of masques and triumphs," the opening sentences of which makes this attitude clear: "These things are but toys, to come amongst such serious observations. But yet, since princes will have such things, it is better they should be graced with elegancy than daubed with cost" (Bacon [1996: 416]).

can take a number of forms and locations: in the break from hermeticism, for example, evident in the unique work of such transitional figures as Paracelsus or, later, John Dee; or this origin is found abiding in the recovery of classical learning and the flowering of humanism; or, within the critique of Aristotelian physics.[31] In still other ways the figure of Francis Bacon is often cited as transformative, particularly for his part in the development of the scientific method said to be so richly on display in the *Novum Organum* and *Sylva Sylvarum*.[32] Rather than attending to these narratives of origin, Harkness looks instead to science already in process in Elizabethan London.[33]

In similar fashion, I understand that the desire to find the point of origin of early modern science is doomed to fail since it is always on the lookout for the first appearance of science against a background of the rest of the culture. Instead, if we collapse foreground and background into something like a perpetual middle ground, we are more likely to realize our goal of discerning early modern science. To this end, I would like to suggest that Bacon's *Device* is a text that can serve very well to demonstrate the historically and socially contingent nature of early modern science. I should stress here that the importance of Bacon's *Device* for this discussion of science and literature in sixteenth- and early seventeenth-century England does not lie in any claim to a privileged status as a point of origin either for science or for literature related to science. Indeed, part of the significance of beginning with Bacon's entertainment is to highlight rather the opposite: that points of origin (of science or of literature) are only mythic in nature.[34] We can no more locate a point of origin for science than we can for literature. But as Bacon's *Device* indicates—as indeed does the *Gesta Grayorum* more generally—in the sixteenth and seventeenth centuries (as today), science was always in the process of emergence. And as current critical discussions of science and early modern culture make clear—and this serves as one of the defining features of much contemporary work that distinguishes it from its forebears—science is not simply born in laboratories (or gardens or museums, for that matter) and then exported out to the culture at large.

[31] For examples of these analyses, see two substantial and representative collections of essays: Vickers (1984) and Pumfrey, Rossi, and Slawinski (1991). Pumfrey *et al.* provides a useful topical bibliography, pp. 293–317.

[32] Very useful overviews of Bacon's contributions to the rise of induction and the "scientific method" can be found in Jardine and Silverthorne (2000), Jardine (1974), and Rossi (1978).

[33] For a helpful discussion of the appropriateness of the term "science" (as opposed to "natural philosophy") as applied to the early modern period, see Harkness's "Note about "science"" (xv–xviii).

[34] One of the most enduring of these creation myths is what we can today perhaps safely call the *myth* of the "Scientific Revolution" itself. Park and Daston (2006) address this matter directly, having deliberately omitted from their work the phrase itself, an omission, they suspect, that "is likely to arouse the most surprise" in readers. But given their embrace of what they call "pell-mell change at every level" (p. 13), this is an omission that is perfectly (if controversially) deliberate:

The cumulative force of the scholarship since the 1980s has been to insert skeptical question marks after every word of this ringing three-word phrase, including the definite article. It is no longer clear that there was any coherent enterprise in the early modern period that can be identified with modern science, or that the transformations in question were as explosive and discontinuous as the analogy with political revolution implies, or that those transformations were unique in intellectual magnitude and cultural significance. [pp. 12–13]

As Bacon's *Device* makes clear, science pre-dates its laboratories and museums that are in fact produced as a response to the existence (perhaps nascent) of science. Laboratories, like gardens and museums, are as much a manifestation of culture as are theaters, voyages, and narratives about them. Much of the most engaging work in early modern science studies—especially by those writers and critics interested in literary culture—has precisely to do with the articulation of the moments and sites of the emergence of science which have multiplied exponentially in recent years. No longer (or even, no longer particularly) imagined as located wholly or exclusively in universities or colleges or private organizations (such as the Royal Society) or the church (such as the *Collegio Romano* in Rome), early modern science is instead seen as emergent within a wide array of cultural practices and cultural sites. Readers and critics today are more likely to understand sixteenth- and seventeenth-century science—and its many and varied literatures—as profoundly local in nature and in circumstance, and therefore as contingent. In a word, they are emergent across the full spectrum of early modern culture and its practices. And the burden of responsibility for readers and critics today is no longer the fissiparous labor of division, of relating literary texts to a contemporaneous scientific background (or the inverse: relating scientific "events" to a literary background), but rather an integrative one: to conceive of what I would like to call culture-science, a unity that predates the subsequent (though problematical) division of the kingdom. This is the special promise of new studies of early modern science-culture for, though never wholly detached from each other, even in the early twenty-first century, the intimacy of science with culture and the unity of their connectedness is more readily recognizable to us than in our own era characterized by the widespread belief (fostered by the politics and, especially, the economics of Big Science) that science has been cordoned off as a special and privileged room of its own.

A parallel concern of this book is not only to consider (with Findlen and Harkness and other historians of science) a number of the multiple sites across the early modern period in which scientific practices are indeed underway, but rather to consider sites—or practices—where what doesn't immediately look like science is underway. It is in this spirit, then, that I wish to consider a range of coordinated sites and practices—the theaters, to name one example, or the early modern sick room, to name another—in which we can indeed observe the work of science in action. In pursuit of these objectives (as I will describe in more detail below) *The Machine in the Text* considers a range of practices—technological and typically prosthetic in nature—that thereby earn the name *machines* and can be seen at work across the period. And even as I wish to move with Harkness beyond conventional definitions and narratives about origins and locations of early modern science, so too do I want to consider the early modern machines in texts as instances of a more general machine imagination at work in the culture—a project indebted to the work of Jonathan Sawday in his recent book *Engines of the Imagination*.[35] But

---

[35] Sawday (2007). See also Wolfe (2004).

there is one major difference: where Sawday wants us to see mechanism and machine where before we saw only the paleotechnic, I want us to see the machine that is less obvious but that nevertheless is emergent and serves in important ways to structure the cultural work at hand.

The objective of this book is *analytically* to pry apart science and non-science, two discourses that in the early modern period were *culturally* imbricated. The *Gesta Grayorum* offers a fine example of this virtual inseparability of the science program it imagines and the literary play it offers: in Bacon's *Device* and Davison's *Proteus*. But the disciplinary approaches to each of these discourses, while yielding powerful instances of disciplinary knowledge, are—so long as they stand separated from one another—insufficient to allow us to see the *Gesta Grayorum* in something like the full density of the cultural work it can be said to enact. What is required, then, is an effort to think the scientific and the literary in the same moment, to read the text with something like a pre-disciplinary understanding of the nature of knowledge in the early modern period.

At times, this can mean reading the literary text with an eye toward its own more or less direct negotiations with the scientific. In *Proteus*, for example, this might mean noticing the masque's concern with the adamantine rock as a glimpse into one way of thinking about the phenomenon of magnetism in the half dozen years prior to William Gilbert's own version, *De Magnete* (1600). Or it may mean paying rather more attention to the masque's engagement with natural knowledge embedded, as it is, within the masque's mythic celebration of Elizabeth. The Squire can thereby deliver his remarkable speech meant to teach Proteus the proper natural order of things in which Elizabeth presides over the world properly ordered by her "attractive" presence. Thus, Proteus's praise of his rock and its likeness to the "Polar star,/Because it drawes the needle to the North" is undermined by the moon (Cynthia), "Whose drawing virtue governes and directs/The flotes, & reflotes of the Ocean," Proteus's reputed domain. But even the moon is displaced by a greater Cynthia, the Prince's greater adamant, Elizabeth:

> But Cynthia praised bee your watry raigne,
> Your Influence in spirits hath no place.
> This Cynthia high doth rule those heavenly tydes,
> Whose Soveraigne grace, as it doth wax or wane
> Affections so & fortunes ebb and flow.
>   . . .
> What excellencies are there in this frame,
> Of all things which her virtue doth not draw:
> The Quintessence of wittes, The fier of loves
> The Ayre of fame, Mettall of courages;
> And by hir virtue long may fixed bee,
> The wheele of fortune and the Car of tyme.
>   . . .
> . . . and Proteus for the seas,
> Whose Empire lardge your praised rock assures,
> Your guift is void, it is already heer,

> As Russia, China, & Magellanus straytes
> Can wittnes beare . . . [*GG*, 84–5]

At other times accessing pre-disciplinary knowledge is a matter of understanding the scientific more in terms of its cultural specificity and immediacy. Thus, as an example, one can read Bacon's *Device* within the context provided by that system of representation and performance that Clifford Geertz calls "the theater state."[36] Having heard the responses of his closest advisors and counselors to the question he had asked them—"of the Scope and End whereunto you think it most for our Honour, and the Happiness of Our State, that Our Government be rightly bent and directed" (*GG*, 44)—the Prince thanks them in a public speech, though he is careful to announce that he will make no rash decision about a course of action: "We should think Our Selves not capable of good Counsel, if, in so great variety of perswading Reasons, we should suddainly resolve." The decision deferred, the evening's cere-monies conclude with "Occasion of Revelling," as Prince and counselors and courtiers choose partners for a final dance (*GG*, 56). The next day, the Prince and his entourage undertake a progress to the Lord Mayor's house: "from *Grays-Inn*, through *Chancery-lane*, *Fleet-street*, so through *Cheap-side*, *Corn-hill*, and to *Cosby's Place*, in *Bishop's-gate-street*." No ceremony is overlooked, nothing left to chance: "This Shew was very stately and orderly performed; the Prince being mounted upon a rich Foot-cloth, the Ambassador likewise riding near him; the Gentlemen attend-ing, with the Prince's Officers, and the Ambassador's Favourites, before; and the other coming behind the Prince; as he set it down in the general Marshalling, in the beginning." The report makes dutiful mention of "the sumptuous and costly Dinner," as well as the "variety of Musick, and all good Entertainment" offered to the Prince and his attendants, but seems eager to tell of the return journey through the streets of the city. With the account of the return progress we find the reason for the report's concern lies with the spectators who assemble to bear witness:

> The Dinner being ended, the Prince and his Company having reveled a while, returned again the same Way, and in the same Order as he went thither, the Streets being thronged and filled with People, to see the Gentlemen as they passed by; who thought there had been some great Prince, in very deed, passing through the City. [*GG*, 57]

The Prince is not ("in very deed") a Prince after all and what the awed and gaping spectators bear witness to is not his royal greatness, but rather their own altogether common and complete ignorance of this fact. The Progress, like the Lord Mayor's dinner, is indeed only a "Shew," a performance carefully modeled on past royal progresses through London and enacted in self-conscious and deliberate imitation of them. And to the extent that royal progresses and similar rituals of display were themselves self-consciously and deliberately theatrical—if they were, as we have come to believe, instantiations of "the illusion of power," in Stephen Orgel's landmark formulation—then what the citizens who had gathered to witness the progress were seeing was, in effect, a second-order theatricality enacted by the

---

[36] Geertz (1980).

"Gentlemen" who can be said to be staging or performing their own thorough mastery of precisely the manipulation of theatrical representation, of both illusion and power, essential to the early modern state—and, of course, to Bacon's celebrated scientific fantasies given such vivid expression in the *New Atlantis*.[37]

Questions about the nature of attending to the scientific and attending to the literary (or the cultural, more generally) are, on the one hand, certainly questions of methodology: the interpretive strategies conventionally deployed by these two critical practices seem to be engendered by the nature of their respective objects of inquiry, objects that we have been pleased to construe as fundamentally different from one another. On the other hand, however, the very distinction between the scientific and the literary is itself a function of culture, and not a matter of "natural fact." It is a distinction, in other words, that has a social history, even if that history has tended to become obscured or even perhaps lost across time. As the example of the *Bulletin of the Atomic Scientists* with which I began this chapter indicates, we are periodically re-called to this knowledge of the connectedness of science and culture at large. Indeed, to look once more to the April 1966 number of the *Bulletin*, we can see an extraordinary discussion in the opening article of the precise nature of science, particularly in regard to its relation to culture, in contemporary attempts to formulate some more or less durable definition of "man" or of "human nature." The essay begins with the following brief consideration of the particular and particularly difficult question of how precisely to define science:

> Toward the end of his recent study of the ideas used by tribal peoples, *La Pensée Sauvage*, the French anthropologist, Lévi-Strauss, remarks that scientific explanation does not consist, as we have been led to imagine, in the reduction of the complex to the simple. Rather, it consists, he says, in a substitution of a complexity more intelligible for one which is less. So far as the study of man is concerned, one may go even farther, I think, and argue that explanation often consists of substituting of complex pictures for simple ones while striving somehow to retain persuasive clarity that went with the simple ones. [p. 93][38]

This is Clifford Geertz writing at the outset of his article "The impact of the concept of culture on the concept of man." The complex picture of "man" that Geertz wants to offer in lieu of the simpler one represented, on the one hand, in the Enlightenment settlement ("natural man") and, on the other, in classical anthropology ("consensual man"), is that figure understood neither as wholly a function of innate capabilities implanted by nature itself, nor merely as the sum total of his actual behaviors. Instead, Geertz focuses on the connection between these two possibilities—between these two figures and, moreover, between the methodologies that propose them:

> Man is to be defined neither by his innate capacities alone, as the Enlightenment sought to do, nor by his actual behaviors alone, as much of contemporary social science seeks to do, but rather by the link between them, by the way in which the first is

[37] Orgel (1975).
[38] Geertz (1966). Geertz's article was also published—in a longer version—in Platt (1965: 93–118); all citations are to this edition.

transformed into the second, his generic potentialities focused into his specific performances. [p. 116]

The complexity Geertz understands to characterize this picture of "man" is in part a function of the methodological necessity we face if we wish "to encounter humanity face to face" since such an encounter requires attention not to types or stereotypes, but rather to individuals: "We must, in short, descend into detail, past the misleading tags, past the metaphysical types, past the empty similarities to grasp firmly the essential character of not only the various cultures but the various sorts of individuals within each culture" (p. 117). On this account, the power of Geertz's new image of man is a function (and an illustration) of the clarity and precision offered by scientific explanation (of the sort described by Lévi-Strauss):

> In this area, the road to the general, to the revelatory, simplicities of science lies through a concern with the particular, the circumstantial, the concrete, but a concern organized and directed in terms of the sort of theoretical analyses that I have touched upon—analyses of physical evolution, of the functioning of the nervous system, of social organization, of psychological process, of cultural patterning, and so on—and, most especially, in terms of the interplay among them. That is to say, the road lies, like any genuine quest, through a terrifying complexity. [pp. 117–18]

Given the concern evident in these passages, it is not too fanciful to suggest that the object of Geertz's analysis is as much science as it is the image of "man" his anthropology affords. This is evident not only in the foregoing descriptions of the general and revelatory simplicities of science, but also in Geertz's sense of how the Enlightenment image of "natural man" emerged from the early practices of sciences in the first place. For it was "natural science, under Bacon's urging and Newton's guidance" that had "discovered" that man was "wholly of a piece with nature." The view of human nature thus produced was marked by its absolute regularity—"as regularly organized, as thoroughly invariant, and as marvelously simple as Newton's universe." Geertz continues: "Perhaps some of its laws are different, but there *are* laws; perhaps some of its immutability is obscured by the trappings of local fashion, but it *is* immutable" (p. 94).

Geertz of course rejects the entire edifice of uniformitarian theories of human nature and the "stratigraphic" methodologies (p. 98) that underwrite them in favor of a more synthetic set of practices that are more apt and more powerful exactly because they are multidisciplinary: "It is a matter of integrating different types of theories and concepts in such a way that one can formulate meaningful propositions embodying findings now sequestered in separate fields of study" (p. 106). These new approaches to the study of human nature allow for the emergence of an understanding of culture as "not just an ornament of human existence but—the principal basis of its specificity—an essential condition for it" (p. 108). And this more synthetic understanding of culture—which is actually an understanding about culture and nature—allows for a more robust definition of "man" ("our ideas, our values, our acts, even our emotions . . . like our nervous system itself") as "manufactured." Men, Geertz declares, "every last one of them, are cultural artifacts" (p. 114).

The fulcrum that leverages Geertz's argument that "there is no such thing as human nature independent of culture" (p. 112) is the insight—gathered from a wide range of scientific and social scientific discourses—that what Geertz calls "cultural patterns," together with the body and the brain, have collaborated across evolutionary time in the construction of a "positive feedback system" in which each shapes the others (p. 111). As a result, nature and culture are so thoroughly collaborate as to deny absolute priority to either (as the "natural man" and the "consensual man" theories attempted to do). For Geertz and the scientific explanation his argument provides, even on the anatomical and the physiological levels, we have always been *cultural* artifacts.

I would like to borrow from this argument this idea of an imbricated nature-culture that by virtue of its overlapping of one with the other provides a useful way to think about the relation between science and the literary (or the cultural more abstractly). To follow Geertz's lead, then, is to suggest that the literary is always scientific and the scientific, by its nature, is literary. This is an argument that the rest of this book will seek to secure against our more traditional patterns of strictly enforced disciplinary separateness that sees a line between the literary and the scientific—between, say, *The Masque of Proteus* and Bacon's *Device*—and, moreover, avers a nearly ontological separation between them. Both of these texts, as records of specific events located—or, indeed, created—collaboratively within a specific time and specific setting, assume a certain oscillation between the literary and the scientific, between the artificial and the natural that we saw at the beginning of this chapter figured in the x-ray image of beryl.

But this scene in which the literary and the scientific are constructed is more complex than this formulation perhaps indicates. The beryl x-ray image, for instance, is a perfectly apt emblem of the *scenario of construction* in which an object that is both natural and artificial is called into being. The diffraction pattern of beryl is a natural effect of an artificial process, the resulting natural object of a set of artificial operations. But for all of that—for all of the x-ray technology and the photographic technology brought into the scenario (to say nothing of the print technology that makes the mass reproduction of the image possible)—what we see when we look at the image is indeed natural. But it is a piece of the natural world, we can say, that becomes possible, that, indeed, can be said to owe its very existence as observable and as observed, only by virtue of the multidisciplinary machinic scenario of its production. It is this scenario—found in a wide array of cultural settings and within an array of undertakings in early modern England and Europe dedicated to the production of knowledge—that will be the principal concern of the following chapters.

3

Hamlet's Machine

Thine evermore, dear lady, whilst this
machine is to him,
Hamlet.

Hamlet, 2.2.122–3

I would like to begin this chapter on *Hamlet*, perception, the nature of experience, and knowledge by revisiting a discussion of the play that has long been influential, even if controversial: T. S. Eliot's 1919 essay, "Hamlet and his problems."[1] Embedded as it is in perhaps excessive thoughts of "that most dangerous type of critic: the critic with a mind which is naturally of the creative order" (p. 121) and Eliot's own vexed relation to both individual talent and tradition, as a critical engagement with *Hamlet* the essay is certainly a failure. In fact, at the risk of a certain repetition or echo—the risk, that is, of a certain *haunting*—I feel compelled to say that probably more people have thought Eliot's essay a work of criticism because they found it interesting, than have found it interesting because it is a work of criticism. It is the *Mona Lisa* of criticism. However, having said this, I hasten to add that Eliot's essay is not without significant interest as a symptom of a certain condition that could be called modernity, but which for the moment I will instead call the loss of experience. Indeed, in "Hamlet and his problems" Eliot seems to have mistaken his topic, since it is clearly more compelling as a discussion of history than of literature. While it fails as literary criticism, in other words, it rather succeeds—though perhaps in spite of itself—as history.

Eliot's determination that *Hamlet* is a failure results mainly from his famous assertion that Hamlet lacks an "objective correlative" that would authorize his emotional state so dramatically and excessively on display throughout the text. Eliot writes:

> The only way of expressing emotion in the form of art is by finding an 'objective correlative'; in other words, a set of objects, a situation, a chain of events which shall be the formula of that *particular* emotion; such that when the external facts, which must terminate in sensory experience, are given, the emotion is immediately evoked. [pp. 124–5]

---

[1] Eliot (1950: 121–6).

Having isolated the play's real problem—"Hamlet's bafflement at the absence of objective equivalent to his feelings is a prolongation of the bafflement of his creator in the face of his artistic problem" (p. 125)—Eliot can only pass his final negative judgment on the play: "In the character Hamlet it is the buffoonery of an emotion which can find no outlet in action; in the dramatist it is the buffoonery of an emotion which he cannot express in art...We must simply admit that here Shakespeare tackled a problem which proved too much for him" (pp. 125–6). What remains implicit but is nevertheless clear is that what Eliot posits as missing from *Hamlet*, its elusive objective correlative, is not a particular object or even a certain situation, but rather some experience that exceeded Shakespeare's ability to understand and then communicate to his audience.

It is of course only the prior assumption that communicable experience is simultaneously the necessary condition and the guarantor of aesthetic accomplishment that allows Eliot to celebrate its absence—not only in *Hamlet* (or the *Mona Lisa*, for that matter), but also in Shakespeare's life, as well. In other words, Eliot's concern with the objective correlative and the formula for apt emotional expression mark the degree to which experience itself is understood to be at risk and in need of vigorous defense. It seems, then, that the objective correlative Eliot can posit only through its absence in *Hamlet* is nothing more than the by-product of his demand for information—about Hamlet's experiences, or even Shakespeare's—that would, after the fact, legitimize the play as art.[2] At the same time, Eliot's argument also makes clear that it is the nature of the senses—the "sensory experience" that Eliot claims must be the terminus of certain external facts—that emerges as the true ground of the aesthetic object's intentional discourse. In contrast to the failed *Hamlet*, Eliot points to *Macbeth* as an example of "one of Shakespeare's more successful tragedies" in which the proper emotion is evoked by the proper alignment of external facts and sensory experience:

> the state of mind of Lady Macbeth walking in her sleep has been communicated to you by a skilful accumulation of imagined sensory impressions; the words of Macbeth on hearing of his wife's death strike us as if, given the sequence of events, these words were automatically released by the last event in the series. [p. 125]

These two concerns taken together—the anxiety occasioned by the perceived loss of experience, on the one hand, and the centrality of "sensory experience" to the aesthetic object, on the other—serve to mark the ground upon which Eliot declares the failure of *Hamlet* and at the same time serve to posit the existence of some unknown (and *by definition* unknowable) experience that thoroughly defeated Shakespeare:

---

[2] The actual target of Eliot's criticism is Gertrude, who is (as Eliot has it) insufficient: "To have heightened the criminality of Gertrude would have been to provide the formula for a totally different emotion in Hamlet; it is just *because* her character is so negative and insignificant that she arouses in Hamlet the feeling which she is incapable of representing" (p. 125). For a powerful critique of this argument, see Rose (1985).

Why he attempted it at all is an insoluble puzzle; under compulsion of what experience he attempted to express the inexpressibly horrible, we cannot ever know. We need a great many facts in his biography; and we should like to know whether, and when, and after or at the same time as what personal experience, he read Montaigne, II.xii, *Apologie de Raimond Sebond*. We should have, finally, to know something which is by hypothesis unknowable, for we assume it to be an experience which, in the manner indicated, exceeded the facts. [p. 126]

The reference here to Montaigne's *Apologie* is revealing partly because it tells us more about Eliot than it does Shakespeare, but also because Montaigne's text explicitly engages the problem of perception and the nature of experience that Eliot seems to take so much for granted. In the *Apologie*, Montaigne's skepticism is in fact predicated upon his understanding of the eminent fallibility of the senses: physically unreliable, subject to deceptions and illusions, hopelessly subjective (and therefore non-confirmable or non-verifiable), the human senses constitute something of an epistemological dead-end:

Our imagination does not apply itself to foreign objects, but is formed through the mediation of the senses; and the senses do not understand a foreign object, but only their own passions; and thus what we imagine and what appears to us are not from the object, but only from the passions and suffering of the senses, which passion and which object are different things; thus he who judges by appearances judges something other than the object. And if you say that the passions of the senses convey to the soul, by resemblance, the quality of the foreign objects, how can the soul and the understanding assure themselves of this resemblance, since they have in themselves no commerce with the foreign objects?[3]

Montaigne's *Apologie* offers his analytical dissection (an anatomy, of sorts) of the kind of failed sensory experience upon which Eliot's dismissal of *Hamlet* altogether depends. From this perspective, it is the profound skepticism located in the *Apologie* that Eliot seems unable to confront, more than it is Hamlet's "problems," and more than it is the "insignificant" Gertrude, who in Eliot's view fails adequately to signify because she is no Lady Macbeth. Indeed, Eliot's refusal (or inability) to engage such skepticism, together with his obsessive concerns with the objective correlative, have the unfortunate consequence of blinding Eliot to the fact that *Hamlet* is itself deeply engaged in its own attempts to recover and secure experience against its anticipated or foreseen loss. Where Montaigne's *Apologie* offers his conclusions about the uselessness of the body as a viable means to access knowledge, *Hamlet* stages precisely the struggle of a figure, Hamlet himself, in the very midst of his attempts to resist such conclusions that seem borne out by virtually the rest of the entire world. And there is a further dynamic that serves to complicate our relation to the play still more (and to alienate a reader such as Eliot from it forever): while Hamlet resists the kind of profound skepticism diagnosed by Montaigne, *Hamlet* embraces it.

---

[3] Montaigne (2003: 161).

At the same time, this narrowness of focus on an aesthetic formula and the place of sensory experience in it (and perhaps this is a narrowness of *conception*) also serves to blind Eliot to the degree to which "Hamlet and his problems" and its anxiety over experience and its potential loss is at least as much a commentary on modernity as it is Shakespeare's drama, or the late Elizabethan era more generally. I would like to extend this observation or assertion, thus: the collapse of the perceptual body staged in *Hamlet* (which I will take up at length below)—and the loss of the possibility of experience altogether that this collapse announces—results in a kind of paralysis that is non-specific in nature (though many readers and critics have identified it as *moral* in nature) which has been construed historically as Hamlet's inability to act. *Hamlet* stands, as it were, at the opening of a long history of generalized paralysis—a history we know as alienation (in its various forms)—that seems to attend upon modernity itself.

One of the great theorists of modernity, and of loss, was Walter Benjamin and his "materialist theory of experience" (to borrow Richard Wolin's phrase) provides a series of powerful insights that I have found helpful in trying to think about both modernity and *Hamlet*, especially in relation to the problem of experience.[4] In his great essay "The storyteller," Benjamin identifies a "new form of communication" that arises as one of the conspicuous products of modernity: information.[5] For Benjamin, information (like the mechanically reproduced work of art) represents not an advancement, but rather the degradation of experience almost to the point of absolute loss—the slide toward what can be posited as the zero-degree state of experience within the postmodern situation analyzed (or perhaps manifested symptomatically) in the work of Baudrillard.[6] In addition to offering his historical materialist critique of the rise of the age of information, Benjamin also situates this fall of communicable experience into its debasement in information within the literary aesthetic trajectory that leads from the story to the novel, that hallmark of modernity that can be traced back to *Don Quixote* and that for Benjamin reaches its greatest articulation *in À la recherche du temps perdu*. Benjamin's analysis of the cult (ure) of information is made possible by virtue of his Marxist sensibility, however idiosyncratic it may be, and his understanding of the alienating effects of modes of production and the rise of the middle class in fully developed capitalism.[7]

"The storyteller" is subtitled "Reflections on the works of Nikolai Leskov," that modern writer who, in Benjamin's judgment, manages to maintain the virtues of the storyteller that have become largely lost in the historical rush to information.[8]

---

[4] This phrase is Richard Wolin's (1982: esp. Ch. 7).

[5] Benjamin (1969: 99).

[6] For a discussion of Baudrillard's works as symptomatic, rather than analytical, see Brown (2004, esp. pp. 1–21).

[7] One could also say that this understanding derives from Benjamin's view of post-World War I Europe and the collapse of the *value* of experience: "For never has experience been contradicted more thoroughly than strategic experience by tactical warfare, economic experience by inflation, bodily experience by mechanical warfare, moral experience by those in power" (Benjamin, 84).

[8] Near the end of the essay, Benjamin cites Paul Valéry's assertion that "artistic observation can attain an almost mystical depth." Benjamin quotes Valéry's discussion of the embroidery work of a textile artist:

I invoke Benjamin's discussion of the storyteller in part to suggest that the drive toward meaning (such as mentioned in connection with Eliot's judgment on *Hamlet* cited above) is itself a manifestation of the collapse of communicable experience and its replacement by information—or, to use Benjamin's language, the substitution of modernity's information (events that come to us today "already . . . shot through with explanation") for the pre-modern narrative and its defining "amplitude that information lacks" (p. 89).[9]

Eliot wants *Hamlet* to explain itself, and though the play will ultimately frustrate this desire, it is not a desire that is unknown to the play. Indeed, part of the richness of *Hamlet* derives from the fact that it stages precisely this moment of transition from pre-modern to modern, from narrative to information. To offer here just one example, the play embeds both the story—the events that befall Hamlet as we see them—and the information into which that story is destined to become corrupted in the anticipated narrative that Horatio agrees to live in order to tell:

> And let me speak to th'yet unknowing world
> How these things came about. So shall you hear
> Of carnal, bloody, and unnatural acts,
> Of accidental judgments, casual slaughters,
> Of deaths put on by cunning and forc'd cause,
> And, in the upshot, purposes mistook
> Fall'n on th'inventors' heads. All this can I
> Truly deliver.[10]

Hamlet offers a sustained investigation into the status of knowledge and its relation to information understood in both a general sense (what one believes he or she

---

The objects on which [artistic observation] falls lose their names. Light and shade form very particular systems, present very individual questions which depend upon no knowledge and are derived from no practice, but get their existence and value exclusively from a certain accord of the soul, the eye, and the hand of someone who was born to perceive them and evoke them in his own inner self. [p. 108]

The *artisanal* quality identified in this passage is centrally important to Benjamin's notion of the fate of storytelling—and art, more generally. Benjamin writes:

With these words, soul, eye, and hand are brought into connection. Interacting with one another, they determine a practice. We are no longer familiar with this practice. The role of the hand in production has become more modest, and the place it filled in storytelling lies waste . . . That old co-ordination of the soul, the eye, and the hand that emerges in Valéry's words is that of the artisan which we encounter wherever the art of storytelling is at home. In fact, one can go on and ask oneself whether the relationship of the storyteller to his material, human life, is not in itself a craftsman's relationship, whether it is not his very task to fashion the raw material of experience, his own and that of other, in a solid, useful, and unique way. [p. 108]

[9] Benjamin writes,

The value of information does not survive the moment in which it was new. It lives only at that moment; it has to surrender to it completely and explain itself to it without losing any time. A story is different. It does not expend itself. It preserves and concentrates its strength and is capable of releasing it even after a long time (p. 90).

[10] Shakespeare (1982: 5.2.384–91). Subsequent references are to this edition of the play.

knows about the world), and in the more particular sense that relates to the senses (the mechanisms through which one collects certain information about the world).

At the same time, I will argue in the following pages that there is another context—a parallel history, in effect—in which to locate *Hamlet*: the gradual consolidation of the culture of science. More particularly, *Hamlet* stages the collision of two narratives. The first of these is the collapse of the perceptual body and its resulting disqualification as a means to knowledge. The second is the emergence of the practices of experimental science deployed in order to recuperate sense perception and re-establish the very possibility of both experience and knowledge.

I

The opening scene of *Hamlet* stages two spectacular and related moments. The first is the anticipated but nevertheless still startling appearance of some "thing"—"this dreaded sight," "this apparition," this "figure like the King" (1.1.24, 28, 31, 44)— that we will learn to call the Ghost of Old Hamlet. The second, which comes about consequentially, is the sudden conversion of the skeptic: "How now, Horatio? You tremble and look pale./Is not this something more than fantasy?" (1.1.56–7). Instantly converted, Horatio replies, "Before my God, I might not this believe/ Without the sensible and true avouch/Of mine own eyes" (1.1.59–61). Typical of this play noted for rendering equivocation an art form, by highlighting the fundamentally important issue of the relation between seeing and knowing that lies at its heart, Horatio's reply serves to obscure precisely those complexities that obtain between the senses and knowledge that the rest of the play' will investigate with concentration and rigor.

For Horatio, the very act of seeing this "illusion" serves definitively to establish the veracity of Marcellus's report. Knowing arises naturally from the body's senses: once one sees a ghost, there simply are no complexities surrounding the relationship between sight and knowledge. Understandings (such as Horatio's) of the ways in which knowledge is made only through the body's senses were underwritten by standard late medieval and early modern models of sense perception. Though these models certainly provided for occasions of the senses leading to confusion or error, they nevertheless insisted upon the viability of the body as a means to knowledge as a natural necessity: the senses provide, after their particular fashions (immediately, for taste and touch, remotely for sight, hearing, smell), stimuli (what we might now call information or data) to the Imagination or Fancy, which then "processes" this input by virtue of an appeal to the instrument of judgment, the Understanding.[11] The relevant literature of the period offers many examples of this understanding of the perceptual system. The following passage from Samson Lennard's 1607 translation of Pierre Charron's *Of Wisdome Three Books* (1594) is both representative

---

[11] Among the many important early modern treatments of the senses and the perceptual system more generally are Laurentius (1599), Wright (1604), and Crooke (1615).

and (due in part to its *textual* figuration of the work of memory) especially interesting to this discussion:

> The imagination first gathereth the kinds and figures of things both present, by service of the five senses, and absent by the benefit of the common sense: afterwards it presenteth them, if it will, to the understanding, which considereth of them, examineth, ruminateth, and judgeth; afterwards it puts them to the safe custodie of the memorie, as a Scrivener to his booke, to the end he may againe, if need shall require, draw them forth (which men commonly call *Reminiscentia*, Remembrance) or else, if it will, it commits them to the memorie before it presents them to the understanding: for to recollect, represent to the understanding, commit unto memorie, and to draw them foorth againe, are all works of the imagination ... [12]

We will return later to the scrivener in the figure of Hamlet and his tables (1.5) and to the matter of memory in general, but for the moment it is important to stress what this passage (and the countless others of its kind) argues: that knowledge is possible only by virtue of the perceptual body. It is also important to mark the historical specificity—and hence the provisional nature—of this settlement: as Bruce Smith reminds us, it would not be until after Descartes in the 1630s and 1640s that we would become convinced that we could "think without [our] bodies."[13]

Hamlet is important to this discussion of the senses in early modern culture in part because it marks something of a crossroads, a moment of the jarring coincidence of two radically opposed epistemologies distinguished by no one feature more than by the different ways in which the role of the body in the production of knowledge is understood. On the one hand, thinking happens only through the body and its properly functioning perceptions. On the other hand, the late sixteenth and early seventeenth centuries witnessed an increasingly serious skepticism over the viability of the body and its perceptions as the mechanisms that secure knowledge. Indeed, to continue briefly to focus on the play's opening scene, even with the support of such theoretical understandings of the nature of knowledge and its dependent relation upon the senses as Pierre Charron's *Of Wisdome* provides, there is something unsettling and disturbing in this staging of the skeptic's conversion and its aftermath. Arguments such as Montainge's *Apologie* mentioned above suggest that Horatio's absolute confidence in his new knowledge may be in fact less durable and less than compelling since it depends wholly upon the testimony provided by sight: he can only establish the "truth" of the apparition through the "true" testimony of his senses. In a word, Horatio knows what he sees because he sees it. This is not to suggest that Horatio does not *believe* what he sees—especially since for him the truth of this experience is registered somatically, his body (pale and trembling) obligingly manifesting the authenticating markers of this truth. But it is to call into question the reliability of an epistemology such as

---

[12] Charron (1607; repr. 1971: 50).
[13] Smith (2004: 149). For other important recent discussions, see Mazzio (2005) and Paster (2004).

Horatio's founded on an uncritical acceptance or understanding of the relation between the senses and knowledge.

For Horatio—as for fifteenth- and sixteenth-century practitioners of natural history, on the one hand, and theorists of the natural (whether humoral or mechanical) body, on the other—seeing is indeed believing precisely because seeing constitutes bodily the means of knowing. At the same time, though, not only is there a generalized philosophical skepticism (of the kind Montaigne embraces and which Horatio seems easily to repudiate), but throughout *Hamlet* the perceptual body that would serve as both the grounds and the means for knowledge fails to function in a determined, known, and reliable fashion. In fact, the perceptual body—that chain of associated apparatuses that begins with the senses and culminates in judgment—is characterized less by a functional reliability (the appearance of an accustomed automaticity of stimulus and response) than by an utter vulnerability, even to the point of collapse. As such, the "true avouch" of the senses trumpeted by the converted Horatio seems decidedly unable to provide any substantial ground for certainty about the world and, moreover, may remain altogether unable to displace a certain skepticism manifest elsewhere in the play. In other words, *Hamlet* asks the urgent question: if thinking happens through the body, then what happens to thought when the body fails? Where Montaigne's *Apologie* offers his conclusions about the uselessness of the body as a viable means to access knowledge (even as he rejects Reason, too, finding faith as the only hope), *Hamlet* stages the struggle to resist such conclusions.

With its profound concern to determine the fate of both knowledge and action in the aftermath of this perceptual collapse, *Hamlet* depicts—and Hamlet inhabits—a strange and unsettling world in which knowledge and experience are figured as impossible because the perceptual body that would serve as their condition of existence simply does not exist. The epistemological crisis the play stages arises from the realization that while thinking cannot happen without the body, it cannot happen with the fallen perceptual body, either. In light of the disqualification of the perceptual body, Hamlet—unevenly, to be sure, and only by fits and starts—attempts to enact a recuperation of knowledge that depends for its success upon a profound recasting of knowledge: no longer understood as the accumulation of meaning that arises naturally from primary perception, this reconstituted knowledge emerges, if at all, rather as the artifact of a deliberate and artificial construction.

In response to the general epistemological crisis inaugurated by the collapse of the perceptual body, Hamlet fantasizes a resolution that aims to recuperate both knowledge and action by means of a strategic new consolidation of experience derived from sense perception. But this strategy can only be realized by *re*-deploying the senses within a sustaining network of practices and techniques that collectively serve to render experience artificial and evidential.[14] And it will be

---

[14] It is perhaps this practice of re-deployment that has been interpreted as repetition or doubling, in various forms, as something of a structural feature of the play. The two most conspicuous and important instances of repetition in the play are the Ghost and the dumb show-Mousetrap scene.

this artificial and evidential experience that in turn will enable the production of an artifactual knowledge capable of leading to action. Hamlet's attempts to secure a way to knowing—figured most powerfully in his turn to theater via *The Mouse-trap*—can be understood, that is, as versions of that particular practice that would eventually become the very hallmark of science: the experiment. For the ambition of the scientific experiment is (as Peter Dear has argued) an artifactual knowledge achieved through the controlled production of artificial experience, as can be seen in such landmark scientific texts of the period as William Gilbert's *De Magnete*, (1600), Galileo's *Sidereus Nuncius* (1610), for example, and William Harvey's *De Motu Cordis* (1628)—companion pieces, as it were, to Shakespeare's play.[15]

In contrast to Horatio as natural historian for whom seeing is sufficient to believing, Hamlet figures as the experimental philosopher for whom seeing is more an occasion for inquiry than it is self-identical to the act of believing. This contrast also describes the differences between the simple empiricists (figured, for example, in the model of natural historians and sixteenth-century antiquarians) and the experimentalists.[16] At the heart of this difference lies the distinction between knowledge construed as merely found, collected, and displayed and knowledge understood as constructed and therefore artifactual. This essential difference is figured in the fundamental distinction between the cabinet of curiosities as the (mere) repository of things, on the one hand, and Boyle's air-pump or Hooke's microscope (to look at but two examples) as machines dedicated to the production of knowledge, on the other.[17]

Hamlet recognizes the radical assault upon the perceptual body that will tend toward its annihilation, even if he cannot understand it: "What piece of work is a man, how noble in reason, how infinite in faculties, in form and moving how express and admirable, in action how like an angel, in apprehension how like a god: the beauty of the world, the paragon of animals—and yet, to me, what is this quintessence of dust?" (2.2.303–08). At the same time, Hamlet will resist this slide toward an inevitable meaninglessness that will prevail once the perceptual body can no longer serve as the means to knowledge, or to action. In the following pages, I will trace the destruction of the perceptual body, enacted by means of acts of

For influential discussions of *Hamlet* and repetition, see Derrida (1994) and Hawkes (1985). For a discussion of repetition and sense perception in *Hamlet*, see Caldwell (1979).

[15] For a powerful discussion of the nature of the scientific experiment as designed to produce artificial experience, see Dear (1995) and (2001, especially Ch. 7).

[16] For a ground-breaking study of early modern science in general, and the use of scientific instruments (or machines) in the production of knowledge, in particular, see Shapin and Schaffer (1985).

[17] I should stress that the distinction between "found knowledge" (figured in objects particularly) and "produced" knowledge should not be construed as either perfectly clear-cut nor as strictly sequential or chronological in nature. Evidence for this can be found quite easily by even a brief survey of the kind of science—or knowledge pursuits—undertaken by the Royal Society through the seventeenth century. The appearance of both forms of knowledge pursuit as figured in Horatio and in Hamlet *in the same text* is also, I would offer, further evidence of the imbricated nature of the natural historical and the natural philosophical.

violence visited directly upon the organs of sense and perception, as well as Hamlet's resistance to its disappearance.

This resistance, I will argue, takes the form of two complementary movements that can be identified by their opposite (though non-oppositional) trajectories. The first is in the direction of "inwardness," the reactive and defensive retreat toward a projected interiority (or subjectivity) that has been the subject of well-known discussions in recent criticism of early modern literature and culture.[18] The second, which I will discuss in greater detail, lies in the direction of what can be called "outwardness": the focused attention that emanates from the introjected self outward toward the material world as the object of organized and sustained epistemological inquiry. If understood as *intellectual* as well as "psychological" trajectories, that is, inwardness and outwardness can be said to construct different kinds of experience and therefore provide different avenues for re-securing a way to knowledge. Taken together, these two trajectories represent yet another important contrast in play during the early modern period: between experimentalists such as Boyle and Hooke who seek knowledge through apparatuses specifically designed to extend the perceptual body, and anti-experimentalist rationalists, such as Descartes and Hobbes, for whom thinking without the body obviates the need to recuperate the perceptual body.[19]

Having identified this contrast between what Bruno Latour has called the "constructivist" and "realist" conceptions of knowledge in the early modern period, I want to stress that these two trajectories should not be construed as delimiting two separate cultural domains.[20] Nor for that matter should the first contrast remarked earlier (between knowledge understood as found and knowledge produced) be thought to describe discrete cultural territories.[21] On the contrary, as discussed in the previous chapter, I want to argue that these trajectories cut across culture in general and are therefore as present in the early modern laboratory (as it will come to be called) as they are in other sites where knowledge is under construction: the library, the study, Gresham College, the printing house, and eventually the Royal

---

[18] See especially Maus (1985); see also Schoenfeldt (1999).

[19] Shapin and Schaffer (1985: 36) quote a famous passage from Thomas Birch's four volume history *The History of the Royal Society of London for the Improving of Natural Knowledge* (1756–7) that helps illustrate Hook's understanding that scientific apparatuses "*enlarged* the senses":

> his design was rather to improve and increase the distinguishing faculties of the senses, not only in order to reduce these things, which are already sensible to our organs unassisted, to number, weight, and measure, but also in order to the inlarging of the limits of their power . . . Because of this, as it inlarges the empire of the senses, so it besieges and straitens the recesses of nature: and the uses of these, well plied, though but by the hands of the common soldier, will in short time force nature to yield even the most inaccessible fortress. [3.364–5]

[20] Latour (1983: 18); Latour offers a powerful reading of Shapin and Schaffer's book; see Ch. 2, esp. pp. 15–35.

[21] It is important to stress that this assertion has a corollary: within the gradual dissolution of Aristotilianism (particularly in natural philosophy and science) as a master narrative, the growing realization of the inadequacy of (unaided) sense perception to the development of natural knowledge no doubt coincided with the rise of experimentalism. Whether experimentalism was strictly compensatory or, on the other hand, one of the causes of the inadequacy of sense perception is another matter—and one that is still very much open to debate.

Society. At the same time, they are present in many other cultural practices in the period, in chorographical and cartographical research, for example, in navigational practices (and thereby in early modern global trade), and also upon the early modern stage. What I suggest here is that these trajectories are best understood as cultural forces at play across the range of cultural sites and practices long before they become the increasingly exclusive properties of those practices and techniques eventually organized (or, disciplined) under the rubric of "science."[22] For his part (to consider just one specific instance), Hamlet's hopeful bid for the restoration of sense and the perceptual body takes the form of a fantasy of what I will call the *autonomic body* dedicated to the production of artificial experience and artifactual knowledge.

II

As literary text and theatrical event, *Hamlet* enacts the fall of the perceptual body in which the organs of perception—eyes, ears, nose, mouth, and skin—are simultaneously the means through which one apprehends the material world and the locus of a profound material vulnerability. In this regard, the play participates in an early modern convention: poets (since Petrarch at least) understood the senses, especially vision, as both invasive and violent, even as other writers—from divines to physiologists to dramatists to anti-theatrical polemicists—were convinced of the power of external stimuli often to overwhelm the body and with it reason and judgment.[23] But where *Hamlet* is less conventional is in its articulation of a fantasy of the *invulnerable* body against which the collapse of the perceptual body is all the more striking, and all the more destabilizing. The projection of this body first appears (though only spectrally, as a ghost) embedded ironically within the most conspicuous and most important instance in the play of the staging of the coincidence of bodily perception and corporeal vulnerability: the story of Old Hamlet's ear into which is fatally poured the "juice of cursed hebenon" (1.5.62).[24] This

---

[22] As a corollary to this notion of outwardness (as I have defined it here) dispersed across widely divergent practices in sixteenth- and seventeenth-century culture, I would stress that it is the exclusive possession of none of them. Nor are its many features, including the experiment, broadly conceived, which I will argue below—by way of an important example from the theater of the period—can be found in any enterprise underwritten by outwardness as an intellectual labor. It is not a matter, then, of finding "scientific" characteristics at play in the theater, but rather of identifying in the theater those intellectual practices that are later appropriated by and absorbed into science.

[23] Within the English sonneteering tradition, for example, with its characteristic emphasis on sight and vision, Sir Philip Sidney's *Astrophil and Stella* offers a profusion of illustrative lines, including Sonnet 42, which in part reads "For though I never see them, but straight ways/My life forgets to nourish languished sprites;/Yet still on me, O eyes, dart down your rays;/And if from majesty of sacred lights,/Oppressing mortal sense, my death proceed,/Wracks triumphs be, which love (high set) doth breed" (Sidney [1989: 169, ll. 9–14]).

[24] Indeed, Old Hamlet's ear itself becomes a figure for the entire state poisoned through the (narrative) treachery of "a brother's hand" (74):

> 'Tis given out that, sleeping in my orchard,
> A serpent stung me—so the whole ear of Denmark

"distilment," as the Ghost has it, is also described as "leperous" (64) due first to its effects on the blood, which "with a sudden vigour it doth posset/And curd" (68–9), and then for its subsequent effects on the skin: "And a most instant tetter bark'd about,/Most lazar-like, with a vile and loathsome crust/All my smooth body" (71–3). Old Hamlet's story of his ear joins two related narratives: the first is the play's concern with the organs of sense under siege and attack from the material world; the second is the image of the *smooth body* imagined as having preceded the attack and that thereafter stands as the lost ideal, as that figure of the whole and solid and unblemished body/self against which Hamlet's own body/self (like a satyr to Old Hamlet's Hyperion) is "sullied." The smoothness of this literally fantastic body is a pure idealization—and an idealization, to be more precise, of the figure of the father. It has many analogues: it is akin to Bakhtin's classical body and is perhaps foundational to Freudian (and certain versions of post-Freudian) psychoanalysis. In either form (and there are perhaps others) this smooth body is not only imagined as closed and impervious to incursions and injuries from an almost naturally hostile world, but it is also quite specifically conceived as non-sexual (or pre-sexual) in nature.

As one example of criticism of *Hamlet* that is to an extent predicated on the fantasy of the smooth body, I will point to Janet Adelman's brilliant reading of the play in general, and of Hamlet's idealization of his dead father, in particular—especially as this idealization functions as a defensive reaction against Gertrude's threateningly sexualized maternal body. Citing both Lucianus's poison—"Thou mixture rank, of midnight weeds collected" [3.2.251]—and the Ghost's story of his murder, Adelman argues that the play enacts "an astonishing transfer of agency from male to female" in which "malevolent power and blame for the murder [of Old Hamlet] tend to pass from Claudius to Gertrude in the deep fantasy of the play" (p. 24) For Adelman, the poison that kills Old Hamlet is (on the level of fantasy) not really Claudius's (any more than Lucianus's is his), but rather Gertrude's, and its

> usurpation on wholesome life derivative not from Claudius's political ambitions but from the rank weeds of Gertrude's body. Its 'mixture rank' merely condenses and localizes the rank mixture that is sexuality itself: hence the subterranean logic by which the effects of Claudius's poison on Old Hamlet's body replicate the effects of venereal disease, covering his smooth body with the lazarlike tetter, the 'vile and loathsome crust' that was one of the diagnostic signs of syphilis.[25]

But that this figure of the perfect, invulnerable, complete, body—the body construed, in a word, as smooth—is nothing more than a fantasy is demonstrated

> Is by a forged process of my death
> Rankly abus'd—but know, thou noble youth,
> The serpent that did sting thy father's life
> Now wears his crown. [1.5.35–40]

This in fact is the first of two forgeries: the second (similarly off-stage) is Hamlet's forgery, sealed with his father's ring, of Claudius's letter to England.

[25] Adelman (1992: 25–6).

everywhere by the play itself, for the smooth body and its organs of sense are *naturally* under siege precisely in order that perception itself should be possible. The ears, for example, are either promiscuously open (Old Hamlet's); deliberately, though ineffectually, closed (as Barnardo characterizes Horatio's: "let us once again assail your ears/That are so fortified against our story" [1.1.34–5]); or simply dead (as are Claudius's—and perhaps Hamlet's, as well—at the play's end: "The ears are senseless that should give us hearing/To tell him his commandment is fulfill'd./ That Rosencrantz and Guildenstern are dead" [5.2.374–6]). Whether understood as opened or closed (and there may well be no actual difference here) the ears are invariably conceived as the site of a certain violence that is both perceptual and epistemological in nature: Hamlet tells Horatio, "I would not hear your enemy say so,/Nor shall you do my ear that violence/To make it truster of your own report/ Against yourself. I know you are no truant" (1.2.170–3); Gertrude begs Hamlet: "O speak to me no more./These words like daggers enter in my ears./No more, sweet Hamlet" (3.4.94–6); Claudius reports to Gertrude on the rumors that greet Laertes upon his return to Denmark after the death of Polonius: he "wants not buzzers to infect his ear/With pestilent speeches of his father's death,/Wherein necessity, of matter beggar'd,/Will nothing stick our person to arraign/In ear and ear" (4.5.90–4); and Hamlet's letter to Horatio announces and warns, "Let the King have the letters I have sent, and repair thou to me with as much speed as thou wouldst fly death. I have words to speak in thine ear will make thee dumb" (4.6.20–3).

All of these instances of the coincidence of the ear and violence, particularly the last, are informed by a belief in the immediacy and involuntary nature of somatic responses to aural impressions, the model for which is established by the Ghost of Old Hamlet and its untold story of purgatorial torments:

> But that I am forbid
> To tell the secrets of my prison-house,
> I could a tale unfold whose lightest word
> Would harrow up thy soul, freeze thy young blood,
> Make thy two eyes like stars start from their spheres,
> Thy knotted and combined locks to part,
> And each particular hair to stand an end
> Like quills upon the fretful porpentine.
> But this eternal blazon must not be
> To ears of flesh and blood. [1.5.13–22]

A destructive violence similarly attends other organs of perception, including the mouth, which is cast as the incontinent site of excess figured both in Claudius's "King's rouse" (1.2.127)—his "heavy-headed revel," as Hamlet calls it (1.4.17)— and later in Hamlet's own emphatic declaration, "Now could I drink hot blood,/ And do such bitter business as the day/Would quake to look on" (3.2.381–3), and culminates in the poisoned drink forced down Claudius's throat by the already dying Hamlet, "Here, thou incestuous, murd'rous, damned Dane,/Drink off this potion. Is thy union here?/Follow my mother" (5.2.330–2).[26] And the nose, passive organ of the sense of smell, is frequently accosted by the bad smells of corruption ("O, my offense is rank, it smells to heaven;/It hath the primal eldest curse upon't—/A brother's murder" [3.3.36–8]) and of death ("But if indeed you find him not within this month, you shall nose him as you go up the stairs into the lobby" [4.3.35–7] and "Dost thou think Alexander looked o' this fashion i'th'earth? . . . And smelt so? Pah!" [5.1.191–2 and 194]).

The general collapse of the organs of sense—or, more abstractly, corporeal sense itself—and the concomitant instability of the smooth body are pervasive and threaten to overthrow all reason, all nature, and even language itself. Here is Hamlet in his remonstrating speech to Gertrude in her bedchamber: "Have you eyes?" he asks her not once but twice in three lines and, in the face of the portraits of her two husbands, expresses his astonishment at the abject failure of Gertrude's senses to perceive and, since she fails to blush at her own error/transgression, the attendant failure of her body to participate in the somatic economy of signification:

> Have you eyes?
> Could you on this fair mountain leave to feed
> And batten on this moor? Ha, have you eyes?
> . . .
> Sense sure you have,
> Else could you not have motion; but sure that sense
> Is apoplex'd, for madness would not err
> Nor sense to ecstasy was ne'er so thrall'd
> But it reserv'd some quantity of choice
> To serve in such a difference. What devil was't
> That thus hath cozen'd you at hoodman-blind?
> Eyes without feeling, feeling without sight,
> Ears without hands or eyes, smelling sans all,
> Or but a sickly part of one true sense
> Could not so mope. O shame, where is thy blush? [3.4.65–81]

As these lines suggest, even after the traumatic and grotesque death of his father, Hamlet maintains as ideal the very notion of the smooth body that has deconstructed all about him. This fantasy not only holds on to the image (or, the ideology) of smoothness, but also to an entirely conventional understanding of the nature and the status of the senses. Such a conventional understanding posits for the body an unmediated relation to the material world figured (speaking in the abstract) in the event and the somatic response to that event manifest in the *natural* reactions of the perceptual body.[27] It is just this natural and inevitable reaction of

---

[26] Or, the mouth is vulnerable as a point of unwarranted entry for the accusation of lying that pierces the body even to its center: "Who calls me villain, breaks my pate across,/Plucks off my beard and blows it in my face,/Tweaks me by the nose, gives me the lie i'th'throat/As deep as to the lungs—who does me this?" (2.2.567–70).

[27] It is also this same faith that informs Hamlet's final invocation of the somatic response:

> You that look pale and tremble at this chance,
> That are but mutes or audience to this act,

the perceptual body that was depicted earlier in the play, in Horatio's account (another repetition) of the somatic effects of the re-apparition of the Ghost:

> a figure like your father
> Armed at point exactly, cap-à-pie,
> Appears before them, and with solemn march
> Goes slow and stately by them; thrice he walk'd
> By their oppress'd and fear-surprised eyes
> Within his truncheon's length, whilst they, distill'd
> Almost to jelly with the act of fear,
> Stand dumb and speak not to him. [1.2.199–206]

As these brief examples suggest, unlike *Hamlet* the play, Hamlet the prince manifests a faith in the notion of a guaranteed relation between event (stimulus) and bodily or somatic response (reaction). But at the same time, *Hamlet* bears the marks of an unmistakable skepticism regarding the somatic economy upon which many of its principal features are constructed, from the Ghost's non-report of its torments to Hamlet's violent repudiation of Gertrude's somatic failures ("I will speak daggers to her, but use none" [3.2.388]). Given the besieged status of the senses, the information provided by them no longer serves to construct a stable basis for action in the world. And it is this destabilization of the senses that accounts for the nostalgia in Hamlet for the smooth body whose somatic economy of stimulus and response is imagined as immediate and effectual.

This crisis—as suggested above—creates two complementary and simultaneously enacted narratives in *Hamlet*. The first (though not the necessarily chronologically or logically prior) is marked both by a concern with and a move toward inwardness, while the second is a corresponding (but opposite) concern with and move toward outwardness. As intended here, outwardness can be understood as the linked move toward the world outside of the (newly conceived) early modern self, a deliberate engagement with the external world through a set of early modern mechanisms and practices that were in the very process of development that have collectively become consolidated under the category "science." For Hamlet, this means the attempt to recuperate experience through the recovery of the functional perceptual body that his world has already disallowed. More particularly, Hamlet seeks to restore perception in the image of the smooth body and it is just this artificial—this constructed—autonomic body that is figured in one of the play's most curious locutions, the figure of the machine found in Hamlet's love letter to Ophelia:

> Had I but time—as this fell sergeant, Death,
> Is strict in his arrest—O, I could tell you—
> But let it be. [5.2.339–43]

This speech serves to align Hamlet with Old Hamlet, neither of whom will be permitted to tell their harrowing tale, and both of whom can be said to leave the matter of re-narration in the hands of another.

> *Thine evermore, most dear lady, whilst this machine is to him,*
> *Hamlet.* [2.2.122–3][28]

At the same time that this line has given readers fits, it has also seemed somehow integral to the greatness of the play—not least because of its indexical nature: the line itself perhaps bears the traces of its original speaker having endured fits (of melancholy or of madness) and at the same moment a certain burden of greatness, whether of station, destiny, or (to invoke a concept for which Hamlet may well stand as an icon) *character* itself. Or perhaps, more appropriately, because of all of these factors. For once we posit something like an interiorized subject (or a theatrical "person") for display upon the stage as an object for our entertainment, for our consideration, and for our judgment, we have then introduced the fundamental crisis of modernity: the necessary condition of subjectivity as a kind of basic—or structural—alienation.[29]

But if this line truly is essential to the play, then what does it mean? Virtually every editor of the play, when offering any aid at all, offers a more or less simple gloss: "machine means body." In his 1982 Arden edition, for example, Harold Jenkins is perfectly typical in this regard: "[. . . *whilst this machine is to him*] while this bodily frame belongs to him." But, also in typical fashion, Jenkins elaborates:

> The Elizabethans thought of nature in general and the human body in particular as a mechanism. The word *machine*, without the prosaic associations it has acquired in a later age, refers admiringly to a complicated structure composed of many parts. [Timothy] Bright, e.g., thinks of the body as an "engine" stirred into action by the soul. Hamlet's phrase is nevertheless tinged with his contempt for corporeal life. [p. 243]

---

[28] It is important to note that this letter (one of several that Hamlet writes through the course of the play) introduces doubt—both implicitly and explicitly—into the scene of (imagined?) love:

> *Doubt thou the stars are fire,*
> *Doubt that the sun doth move,*
> *Doubt truth to be a liar,*
> *But never doubt I love.* [2.2.115–18]

There is hardly a more equivocal passage in the entire play, nor one more perfectly illustrative of Hamlet's—and *Hamlet's*—essential ambiguities. The first two lines invite or allow Ophelia to doubt the truth of two certainties: that stars are fire and that the sun circles the earth. Hamlet will allow even these two facts of nature to pass away, provided that Ophelia does not come to doubt his love. The third line, however, disrupts the logic of the opening two: here, *doubt* carries the meaning of *suspect*, and the rest of the phrase violates the parallelisms of the first two lines, so that suspicion would collapse the difference between two opposites. The relation in the third line is no longer between facts and anyone's interest in believing them or not, but rather between truth and lies. The *coup de grâce* is delivered in the final line, in which doubt carries simultaneously both meanings: on the one hand, Ophelia should disbelieve the cold hard facts of the natural world before she should ever disbelieve Hamlet's love; on the other hand, she should be sure never to suspect *that* he loves. And of course, these two contradictory commands are not entirely contradictory, since the *facts* of the natural world, especially the motion of the sun, were very particularly under revision during the time in which Shakespeare writes this play. We cannot expect, within the fictional world of the play, that Ophelia had any doubts about the sun's journey around the earth, but such a doubt was certainly available to members of Shakespeare's audience.

[29] For a discussion of the idea of a "theatrical person," see Dawson (2001).

I must confess that I do not know how to understand this. If in 1600 "machine" had yet to carry "prosaic" associations that would attach to it in a later—an unspecified—age but referred instead in something like wonder to the body as a marvelous mechanism, then the same word or idea cannot properly be said to be "tinged" with Hamlet's contempt for corporeal life, a contempt (indeed, like its opposite, his admiration) that is undeniably evident elsewhere in the play. At least this cannot be the case for Hamlet as he writes these words and so is perhaps more true of Jenkins, or any reader in (we can name it now, perhaps) the post-industrial world in which the machine may well be nothing more than prosaic in association.

The (editorial) equation of "machine" and "body" may seem an unremarkable determination, not only because it makes perfect sense as a lover's vow—"Hamlet is Ophelia's so long as his body endures"—but also because the gloss appears to be guaranteed by the prestige of its genealogical descent from no less a figure than Descartes and the mechanical philosophy he comes to represent. Yet, the substitution of "machine" for "body" strikes me as a particularly unhelpful instance of defining one unclear term (machine) with a second that is vastly *more* unclear and ambiguous (body). Needless to say, whole philosophies and entire religions (to say nothing of certain mystical traditions and a significant number of particular sciences) have been devoted to the discovery, or the invention, of a definition of that most elusive term. And of all Shakespeare's plays, the one *least* likely to yield a robust definition of "body" is probably *Hamlet*, a play that reads very much like a sustained indictment of a body that everywhere manifests the unmistakable signs of its utter vulnerability, even to the point of death and the material dissolution rather fetishistically imagined literally in and beyond the grave: "To what base uses we may return, Horatio! Why, may not imagination trace the noble dust of Alexander till a find it stopping a bung-hole?" (5.1.196–8). By these measures, if the body is a machine, it is clearly a very poor one indeed.

But surely this cannot be the whole import of Shakespeare's only use of the term machine; "There needs no ghost," as Horatio says, "come from the grave/To tell us this" (1.5.131–2). But "machine=body" is only one possibility and it would be just as reasonable to suggest that the "tinge" of machine Jenkins notes—and that is perhaps evident in *Hamlet*—is embedded within the very notion of the machine itself (prior to its prosaic fate not yet experienced by Shakespeare in his age but duly noted by Jenkins in his), but inherent in machine-ness at all times. In other words, perhaps the "tinge" of the machine, like the wonder and admiration it can also inspire ("What piece of work is a man"), is fundamental to the fact of the machine from the beginning.

On this reading, then, Hamlet's machine constitutes a problem. One way out of this difficulty is suggested by recent work in science studies—a term meant to include a variety of related disciplines and practices (the history, sociology, philosophy of science, among them). The interpretive havoc wrought by the more or less unselfconscious act of saying that "body" is the real-world substitute for "machine" in Hamlet's famous line can perhaps be undone by, in a sense, reversing the polarity of the critical formulation. The question, as I see it, is not: "What is a body?" but instead—or rather—*first*: "What is a machine?"

Questions about the nature and the status of the machine in early modern culture have recently come under serious study—especially in Jessica Wolfe's *Humanism, Machinery, and Renaissance Literature* (2004) and in Jonathan Sawday's *Engines of the Imagination: Renaissance culture and the rise of the machine* (2007).[30] In these important books, both Wolfe and Sawday provide new insights into the complicated ways in which early modern machines were conceived, crafted, used, and—most importantly, but also most challengingly—understood in the period. Both scholars allow not only our recovery of the astonishing ubiquity of technology and machinery in an age too frequently construed in conventional historical accounts as simply pre-technological, but also enable our recovery of the symbolic, philosophical, and moral dimensions of the machine across early modern English and European culture. For Wolfe, whose study is focused largely on the sixteenth century (and the "ways in which Renaissance writers understand machinery and the practice of mechanics before the rise of the mechanical philosophy" [p. 1]), the history she outlines is fundamentally the history of a certain struggle. Within the context of "a revolution in method taking place in natural philosophy [and] in logic, laws, and politics," for humanists "the project of establishing correspondence between mechanical and intellectual methods proves perilous." She continues:

> Humanism and its machines are thus engaged in a series of ongoing struggles, for as strenuously as Renaissance writers attempt to accommodate machinery to their habits of thought, machinery resists that accommodation, alternately exalting and annulling the culture's most cherished values and its most precarious beliefs. [pp. 5–6]

It is this "radical polyvalence of machines in the pre-machine age" (p. 241) that constituted such a challenge to early modern writers, even as it has been a central challenge facing critics and historians attempting to understand the early modern negotiation of what Wolfe calls "the precarious distinction between human and mechanical instrumentality" (p. 114).

Sawday notes in his comprehensive account of the rise of the machine in early modernity that one especially charged dimension within which this precarious distinction between the human and the mechanical (or artificial) instrumentality of the machine was contested drew much of its complexity from the shared theoretical accounts of the processes (and perhaps origins) of the mechanical, on the one hand, and the aesthetic (or literary) on the other:

> Constructing the artificial device of a machine or engine . . . bore striking affinities to the mysterious process by which the equally artificial device of a poem came into being. For poems and machines were both products of *Technē*. Indeed, the language of machines and Renaissance poetic theory seem to slide seamlessly into one another. Many of the terms applicable to Renaissance machinery, such as "device", "contrivance", and "invention", were also words used to describe the artful effects achieved by the poet or the playwright. [p. 174]

---

[30] Wolfe (2004); Sawday (2007). Subsequent references appear parenthetically.

To Wolfe's and Sawday's accounts I would like to add a further borrowing. Following the lead of a number of critics working in science studies, including Bruno Latour on the practices of science and Peter Dear on the nature of the scientific experiment, I propose to think about the machine as an amalgamation of instruments dedicated to the production of artificial experience that is deployed as a machination.[31] If we think of Hamlet's invocation of the machine as a reference to a set of machinations and to a prosthetic system—not of glass vials or mechanical devices but rather of literary and rhetorical uses of such representational systems as enclosed plays, posthumous life stories, and secret forgeries designed to produce artificial experience—then what emerges is a new way to understand both the machine and the body, on the one hand, and Hamlet's attempt to reinvent the possibility of experience and knowledge, on the other. Hamlet's love letter to Ophelia (which I take to have been composed at some point after Act One), emerges as part of a larger plot or ploy to achieve Hamlet's greater objective: not winning or maintaining the love of Ophelia, but satisfying the command of the Ghost to avenge Old Hamlet's murder. If Hamlet's love letter is as much machination as machine it would then be just like his other letters—and indeed, other writings—through the course of the play. His forged royal commission at sea, for example, his disturbing and obscure letter to Claudius ("High and mighty, you shall know I am set naked on your kingdom" [4.7.42–3]), and his staging of "The Mousetrap" ("The play's the thing/Wherein I'll catch the conscience of the King" [2.2.600–1]) are equally instances of machinations that we can call textual in nature and that (like the love letter) employ certain strategies of ambiguity in pursuit of his revenge. These machines are all dedicated to the production of artificial experience and the resecuring of knowledge that experience allows.[32]

If we are serious in considering the machines/machinations as briefly outlined here, then perhaps we can offer for consideration a modified Hamlet, a character for whom the dread command of the Ghost gives rise not to instant action, but rather to something more like multiple, local, and *improvisational* machinations that collectively articulate a general trajectory toward a resolution which in fact is never clearly in sight, and that perhaps changes in part as the result of the machinations and their immediate and sometimes contradictory outcomes. The most important of Hamlet's machines/machinations addresses the collapse of the perceptual body and at the same time offers a resolution to this most basic of

---

[31] Latour writes, "A machine, as its name implies, is first of all, a machination, a stratagem, a kind of cunning, where borrowed forces keep one another in check so that none can fly apart" (1987: 129). For a powerful consideration of the experiment and artificial experience, see Dear (2001).

[32] Perhaps a word of caution is required at this point. Having suggested that Hamlet deploys certain strategies—in his relation to Ophelia and Gertrude, and in his conflict with Claudius—that function as machinations intended to aid him in the realization of the revenge he is prompted to "by heaven and by Hell" (2.2.580), I want to be careful not to give the impression that Hamlet pursues his revenge in anything like a systematic fashion. Indeed, this has long been understood as one of the issues at the very heart of the play: does Hamlet pursue revenge, or is revenge something he rather falls into in the play's climactic moment? Answers to this question have given rise to the more or less conventional reading of the play in which Hamlet is seen, if not as a failed, then at least as a reluctant avenger.

problems: the *autonomic* body itself emerges in the play as the principal machination within Hamlet's projected regime of problem-solving.

Upon the instant of hearing from the Ghost the fact of his father's murder, Hamlet envisions instant revenge: "Haste me to know't." But the very language he uses to express the instantaneous nature of the revenge that arises from him as if it were itself a kind of somatic reaction to a physical stimulus too great or too demanding to suffer even the slightest mediation, betrays Hamlet's predicament: "Haste me to know't, that I with wings as swift/As meditation or the thoughts of love/May sweep to my revenge" (1.5.29–31).[33] What the remainder of the play substantiates is that neither meditation nor the thoughts of love are swift, and through the course of the play both serve to frustrate Hamlet's execution of swift revenge. In fact, precisely the interval between event (the news of Old Hamlet's murder at the hands of Claudius) and the response (Hamlet's moment to kill the King) serves to separate Hamlet from his revenge, in particular, and action, in general. Given Hamlet's alacrity in thrusting his blade through the arras—'O what a rash and bloody deed is this!' (3.4.27)—or his swiftness in securing the execution of Rosencrantz and Guildenstern, he is clearly capable of "sweeping" to action. What intervenes, significantly, is precisely meditation across time, so that the execution of action, as the Player King's speech makes clear, emerges not from valor or from duty, but if at all only *as a function of memory*:

> Purpose is but the slave to memory,
> Of violent birth, but poor validity,
> Which now, the fruit unripe, sticks on the tree,
> But fall unshaken when they mellow be. [3.2.183–6]

Perhaps conventionally understood as a figure for Old Hamlet, in terms of this articulation of "purpose," the Player King is clearly as much a figure of Hamlet and the loss not only of a father, but of what Hamlet had earlier called "the motive and the cue for passion" (2.2.555). The Player King offers something of a parable of Hamlet's predicament: "What to ourselves in passion we propose,/The passion ending, doth the purpose lose" (3.2.188–9). Indeed, the Player King's diagnosis of the interval between passion and action stands as a cautionary tale warning of the

---

[33] The passage quoted here follows from the language of Q2; both Q1 and F provide slightly different versions—and both share the "erasure" of the "I" at the center of Hamlet's proclamation. F reads,

> Haste, haste me to know it, that with wings as swift
> As meditation or the thoughts of love
> May sweep to my revenge.

Greenblatt (2001) comments on this elision:

> Meditation and love figure the spectacular rapidity of thought, not only the virtually instantaneous leap of the mind from here to the moon but that leap *intensified* by the soul's passionate longing for God or for the beloved. It is as if the desire for haste is so intense that it erases the very person who does the desiring: the subject of the wish has literally vanished from the sentence. Yet the metaphors Hamlet uses have the strange effect of inadvertently introducing some subjective resistance into the desired immediacy, since meditation and love are experiences that are inward, extended, and prolonged experiences at a far remove from the sudden, decisive, murderous action that he wishes to invoke. [pp. 207–8]

dangers inherent in the very structure of early modern conceptions of the body's perceptual system and the relation, in particular, of memory to the senses and their stimuli, on the one hand, and the mechanisms of the Understanding (sometimes, the Imagination or Fancy) in the processing of memories, on the other. As we saw in the passage from Lennard's translation of Charron's *Of Wisdome Three Books*, the information (to use that modern term) collected by the senses would either be processed immediately and then routed to "the safe custodie of memorie," or it could be stored in the memory immediately, literally to be recalled and acted upon at some later time.

One significant threat, then, to "purpose" is precisely the interval of time created by the immediate inscription of stimuli as memory and their recollection at some later moment, should passion wane. A second threat to purpose—and therefore an incentive to inaction—lies within the very nature of memory conceived as a (textual) retrieval system in which any particular piece of inscribed information already stands at one significant remove from the original event/stimulus itself. This kind of inscription is always and immediately only a *copy* of a lost ephemeral event that can only be indexed by an inscribed trace within the perception-memory apparatus that is by its very nature impermanent and fluid. Hobbes's famous formulation of memory as "decaying sense" is an especially apt characterization of such fragility:

> And any object being removed from our eyes, though the impression it made in us remain; yet other objects more present succeeding, and working on us, the Imagination of the past is obscured, and made weak; as the voyce of a man is in the noyse of the day. From whence it followeth, that the longer the time is, after the sight, or Sense of any object, the weaker is the Imagination. For the continuall change of mans body, destroyes in time the part which in sense were moved ... This *decaying sense*, when wee would express the thing it self, (I mean *fancy* it selfe,) wee call *Imagination*, as I said before: But when we would express the *decay*, and signifie that the Sense is fading, old, and past, it is called *Memory*.[34]

Hamlet seems aware of the textual nature of memory, as well as the threats posed to it by time and "decaying sense" occasioned by what Hobbes called "the continuall change of mans body." We see the first glimpse of this awareness—and the corresponding anxiety of a virtually inevitable forgetfulness and inaction that it conjures—when Hamlet responds to the Ghost's call to remembrance:

> Remember thee?
> Ay, thou poor ghost, whiles memory holds a seat
> In this distracted globe. Remember thee?
> Yea, from the table of my memory
> I'll wipe away all trivial fond records,
> All saws of books, all forms, all pressures past
> That youth and observation copied there,
> And thy commandment all alone shall live

[34] Hobbes (1996: 16).

Within the book and volume of my brain,
Unmix'd with baser matter. [1.5.95–104]³⁵

The "baser matter" against which Hamlet hopes to insulate the import of the Ghost's tale of murder and its call to revenge is, perhaps surprisingly, not only the "continuall change of mans body," but also thought itself. For Hamlet, who is known to criticism as Shakespeare's most intellectual figure, intellection comes as something of a burden. It is this irony that gives rise to Hamlet's critique of thought for its deleterious effects on resolution:

Thus conscience does make cowards of us all,
And thus the native hue of resolution
Is sicklied o'er with the pale cast of thought,
And enterprises of great pitch and moment
With this regard their currents turn awry
And lose the name of action. [3.1.83–8]³⁶

The hoped-for antidote to the loss of action that is built into the very structure of perception, memory, and action is nothing less than the desire to reinvent memory itself as a physical sense, a mechanism that could be incorporated literally into the body ("incorps'd," to use one of the play's more curious figures [4.7.86]), rather than remain a faculty of sense that separates thought from action. It is Hamlet's own nostalgia for the smooth body, in which response follows in unmediated fashion from stimulus, that generates the fantasy of something that (at least theoretically) is solid in nature and guaranteed in both its reliability and its durability.³⁷ What Hamlet imagines as sufficient to stand in for the absent smooth body of the father is the autonomic body of the machine. It is perhaps this glimpse

---

³⁵ Criticism has led to consideration of what can be called the *textuality of memory* exemplified by this very speech in *Hamlet* (to specify one especially powerful and influential site). This work has been achieved by means of two analytical or interpretive paths that should be considered complementary in nature: criticism of a decidedly deconstructionist bent—illustrated (for example) by the work of Garber (1987, esp. Ch. 6) and Goldberg (2003)—and a more explicitly historicist work on "the book of memory" as represented, for example, by Stallybrass (2001) and Stallybrass, Chartier, Mowery, and Wolfe (2004).

³⁶ This is also the theme of Hamlet's final soliloquy, which in part reads

Now whether it be
Bestial oblivion, or some craven scruple
Of thinking too precisely on th'event—
A thought which, quarter'd, hath but one part wisdom
And ever three parts coward—I do not know
Why yet I live to say this thing's to do,
Sith I have cause, and will, and strength, and means
To do't. [4.4.39–46]

³⁷ It is perhaps worth suggesting here that this fantasy of the autonomic body not only informs *Hamlet* (and Hamlet), but also can be said to underwrite Eliot's fantasy of the objective correlative—that "set of objects," as quoted earlier in this discussion, "a situation, a chain of events which shall be the formula of that particular emotion; such that when the external facts, which must terminate in sensory experience, are given, the emotion is immediately evoked" (pp. 124–5). Or perhaps Eliot's objective correlative *is* the autonomic body: "the words of Macbeth on hearing of his wife's death strike us as if, given the sequence of events, these words were automatically released by the last event in the series" (p. 125).

of the reconstituted smooth body that provides some measure of refuge from the "slings and arrows of outrageous fortune," since (as the rest of that famous speech suggests) death offers no rescue or release. And indeed it is Hamlet's constant flirtation with death or physical dissolution (the "sullied flesh," for example, that he dreams may "Thaw and resolve itself into a dew" [1.2.129]) and his determined rejection of "self-slaughter" (132) that have served to position him between two equally unacceptable alternatives: the inaccessible (because lost) smooth body, on the one hand, and death, on the other.

In the face of this impasse, Hamlet discovers a third alternative: the autonomic body enabled by means of the relocation of memory within an artificial system of stimulus and response that will be made to stand in for the lost perceptual body. This artificial system takes the form of theater and functions, I would like to suggest, as a machine. I refer to Sawday's useful definition of "machine" with which he opens his book. "Just like our own machines," he begins, "Renaissance machines were useful devices with which people worked and laboured." He continues, offering what strikes me as a perfectly apt description of the early modern theaters as machines:

> Acting upon the world, their avowed purpose was to make human existence more tolerable. But fabricated as they were out of a synthesis of poetry, architecture, philosophy, antiquarianism, and theology, as well as craft, skill, and design, Renaissance machines were also freighted with myth, legend, and symbolism. [p. 1]

The notion of the early modern theater as a machine represents one of the dimensions in which the practices of early modern culture emerge differently to our view from the general perspective afforded by new studies of the cultures of science—new studies that (in part) allow us to suspend an excessively strict delineation of practices by the application of excessively strict proprietary understandings of disciplinarity and the pursuits of knowledges. In his recent essay "Life science: Rude mechanicals, human mortals, posthuman Shakespeare," for example, Henry S. Turner cites the new understanding of "science" and the "ways in which it may be closer to the poetry and drama of the early modern period than we might imagine."[38] Arguing that a clear-eyed consideration of the work actually conducted by scientists in the modern laboratories would yield the realization of "how tentative, piecemeal, frustrated, hopeful—in a word, *hypothetical* science really is," Turner highlights the locus of this work "at a threshold between fact and fiction that is genuinely exciting to scientists themselves and that should be equally exciting to literary critics" (p. 197).

Turner's special focus in this essay is the problem represented by the dramatic character. Referencing Pope's famous (if perhaps obscure) assertion that "every single character in *Shakespear* is as much an individual as those in Life itself," Turner will ask, "in what *way* is a dramatic character 'alive'? *How* it is alive, and what can its putative 'livingness' tell us about the concept-in-life in general, and not

---

[38] Turner (2009: 197).

simply about 'human' life"? (p. 203).[39] As these few lines suggest, Turner's argument requires a re-theorization of the very notion of imitation—both as it relates to the idea of the actor who labors to imitate life on the stage, and to the role imitation plays in the work of science and to "scientific epistemologies of life, not merely in sociological or anthropological accounts of life but in the 'hard' sciences of biology and genetics" (p. 203).[40]

As Turner astutely points out, for early modern writers, work of all kinds enacted on this threshold that is "the 'fictive,' or the 'imaginary' and 'invented'" was often located "within the larger problem of the relationship between 'art and nature,' which itself formed the discursive domain for many arguments that we would today describe as 'scientific' or 'technological.'" It is for these reasons, then, that Turner proposes "that we approach the early modern theater as a kind of machine" (p. 204). And although his focus will be a particular use of that theater-machine ("to fashion or to project artificial life" [p. 204]) in the form of the dramatic character, it seems to me that the notion of the theater as a machine has broad appeal and expansive analytical or interpretive potential. And this potential does not go unnoticed by Hamlet.

The Player's Hecuba speech, in 2.2, offers Hamlet his first opportunity to witness the simulated re-emergence not of the smooth body, but of its double, its simulation:

> Is it not monstrous that this player here,
> But in a fiction, in a dream of passion,
> Could force his soul so to his own conceit
> That from her working all his visage wann'd,
> Tears from his eyes, distraction in his aspect,
> A broken voice, and his whole function suiting
> With forms to his conceit? And all for nothing!
> For Hecuba! [2.2.545–52]

What so astounds Hamlet ("What's Hecuba to him, or he to her,/That he should weep for her?" [2.2.553–4]) is that the Player has just demonstrated a solution to both the crisis of the fallen perceptual body and to the threats posed to purpose by the deferral that lies at the heart of memory and that blocks access to action. For the Player can move to action ("his whole function suiting/With forms to his conceit") because he can produce passion artificially. And, student of mimesis that he is (Hamlet is, after all, a classicist, and his theory of art a correspondingly traditional one), he sees the Player as a figure for himself:

> What would he do
> Had he the motive and the cue for passion
> That I have? He would drown the stage with tears,

[39] Turner quotes Pope (1723–5: 1: xxii).
[40] This general concern with imitation in both Shakespeare's *A Midsummer Night's Dream* and in the hard science of genetics is the topic of Turner (2007)—a provocative and exemplary instance of the kind of exciting work that a deep engagement with science studies makes possible in contemporary criticism of Shakespeare and early modern culture.

And cleave the general ear with horrid speech,
Make mad the guilty and appal the free,
Confound the ignorant, and amaze indeed
The very faculties of eyes and ears. (2.2.554–60)[41]

The Player's Hecuba speech, then, triggers a turn to theater that comes as Hamlet's hopeful bid to restore the smooth body and reinvent memory by incorporating it within the newly conceived autonomic body.

But there is something still more particular about Hamlet's turn to theater that serves to make it especially astonishing. Hamlet's image of the utterly amazed body, its very faculties of sight and hearing wholly overcome, provides a hint (both to Hamlet and to his auditors). Hamlet's appeal to theater emerges from his knowledge of an old and trusted axiom (to which this whole speech has been tending) that theater can conjure the guilty to confess: "Hum—I have heard/That guilty creatures sitting at a play/Have, by the very cunning of the scene,/Been struck so to the soul that presently/They have proclaim'd their malefactions" (2.2.584–8). What makes this turn to theater so remarkable is that it is undertaken in the spirit of what will become the very hallmark of science: the experiment. Hamlet's turn to theater is an experiment in both a general or abstract sense, and in a particular and local sense. On the one hand, *The Mousetrap* is an experiment that will test and assess the truth value of the axiom about theatrical representation and the confessions of the guilty, while on the other hand it will test Claudius by trying to cause the artificial experience of an authentic reaction—to fret Claudius (as Hamlet says after the event) with false fire:

I'll have these players
Play something like murder of my father
Before mine uncle. I'll observe his looks;
I'll tent him to the quick. If a do blench,
I'll know my course. [2.2.590–4]

In both of these ways *The Mousetrap* functions as experiment and as such marks one striking moment in that monumental change in the seventeenth century remarked by Peter Dear as central to the emergence of experimental science. "A knowledge of past events," Dear writes, "was not true knowledge; a knowledge of the current state of affairs was itself mere history." He continues:

The question "Why?" in the sense of Aristotle's "Why thus and not otherwise?"—expecting the answer "because it cannot be otherwise"—haunted would-be knowers, heirs to the Western philosophical tradition. "Experience" was understood as a field from which knowledge was constructed, rather than a resource for acquiring knowledge, because "experience" was itself incapable of explaining the necessity of those things to which it afforded witness. By the end of the seventeenth century,

[41] And yet, even this line of meditation leads Hamlet to still more thought, more language, and more words: "Why, what an ass am I! This is most brave,/That I, the son of a dear father murder'd,/Prompted to my revenge by heaven and hell,/Must like a whore unpack my heart with words/And fall a-cursing like a very drab,/A scullion! Fie upon't! Foh!/About, my brains" (2.2.578–84).

however, a new kind of experience had become available to European philosophers: the experiment.[42]

Hamlet resides at the transition between two systems of knowledge: between knowledge understood as evident and crystallized in axioms (in *Hamlet*, the power of theater to elicit confessions), and knowledge produced artificially (in *Hamlet*, Claudius's "blench" and his cry, "Give me some light" [3.2.263]); between an understanding of "how things happen" and "how something had happened."[43] Dear describes this history in more detail:

> At the beginning of the seventeenth century, a scientific "experience" was not an "experiment" in the sense of a historically reported experiential event. Instead, it was a statement about the world that, although known to be true thanks to the senses, did not rest on a historically specifiable instance—it was a statement such as "Heavy bodies fall" . . . By the end of the seventeenth, by contrast, it had become routine, especially in English natural philosophy, to support a knowledge-claim by detailing a historical episode. (pp. 13–14)[44]

Claudius's revulsion at the re-enacted scene of his fratricidal murder of Old Hamlet (itself a re-enactment of the first murder—"It hath the primal eldest curse upon't" [3.3.37]), is a somatic response, which Hamlet will take as a marker of truth, that is *only* produced artificially. For Hamlet, Claudius's "occulted guilt" manifested itself somatically: the "blench" (or flinch) Hamlet witnessed arose naturally from the artificial simulation of the king's guilt that he had been made to see: "If his occulted guilt/Do not itself unkennel in one speech,/It is a damned ghost that we have seen,/And my imaginations are as foul/As Vulcan's stithy" (3.2.80–4).[45] Claudius's guilt becomes Hamlet's knowledge because it is produced *automatically*, as a matter of course or nature, and because it was made to do so without the mediatory roles of speech ("For murder, though it have no tongue, will speak/With most miraculous organ" [2.2.589–90]) and—even more importantly—without thought.[46] *Hamlet* manifests a certain understanding of that defining feature that will come to characterize the modern—and the scientific—world: that meaning, like authenticity,

---

[42] Dear (1995: 11–12).

[43] Dear (1995: 4).

[44] Dear's work is important in helping us to understand this transition, especially its relation to the experiment: "The new scientific experience of the seventeenth century established its legitimacy by rendering credible its historical reports of events, often citing witnesses. The singular experience could not be *evident*, but it could provide *evidence*" (1995: 25).

[45] Hamlet's speech continues, and as it does, it makes clear the degree to which the observation of the somatic response is itself vulnerable to mere performance: "Give him heedful note;/For I mine eyes will rivet to his face,/And after we will both our judgments join/In censure of his seeming" (3.2.84–7). The obvious comparison is to Hamlet's dismissal of "actions that a man might play" in Act One: 'Seems, madam? Nay, it is. I know not 'seems'" (1.2.76).

[46] And this is both the attraction and the dread of the autonomic body: it functions with a machinic reliability and regularity of pure mechanism, but at the same time perhaps threatens to (d)evolve into the soul-less automaton that haunts the emergent culture of science in the period, from the anatomical researches that succeed in demonstrating physical structures (the fabric of the human body) but fail to locate life's animating force, to the mechanical philosophy of Descartes and Hobbes, to the long-lived obsession with the fabrication of actual automata in ventures as diverse as garden architecture and the construction of automatic musical instruments.

is not a naturally occurring fact, but rather the artifact of a complicated set of technical operations.

If Hamlet functions (at least in part) as both reader and interpreter of this set of technical operations, then the term that can perhaps be used to describe Hamlet as the master of these technical operations itself emerges (one could say) from the figure of the machine or the engine: *engineer*. As Jonathan Sawday reminds us, the term/concept *engineer* "existed in the sixteenth century in a different and rather richer sense than it does today" and it is a term that rather aptly names the figure in which Hamlet casts himself when contemplating his next strategy (his next machination) in the face of the duplicity of Rosencrantz and Guildenstern. In order to thwart their efforts, Hamlet "deploys a cunning and artful 'device' or plot, which is conceived of in terms of the devious skills of the military engineer" (p. 99). The reference is to Hamlet's awareness that he can in no way trust his one-time friends—and at the same time to Hamlet's representation of what Sawday calls his "calculations" (p. 99): "Let it work;/For 'tis the sport to have the enginer/Hoist with his own petard, and 't shall go hard/But I will delve one yard below their mines/And blow them at the moon" (3.4.207–11).

Hamlet's plan at this moment in the play cannot be a specific one. If what he later writes to Horatio about his midnight discovery at sea of Claudius's secret warrant commanding Hamlet's death and then his own forged commission with which he replaces it is true, he does not yet (in the closet scene) know anything more than that he is to be sent to England and that his former friends are a danger to him because they are indeed the king's loyal subjects. But for all of that, Hamlet knows that he plans to subvert the efforts of Rosencrantz and Guildenstern, though the devising and the execution of this plan emerge entirely as contingent improvisations, part inspiration, perhaps, and part opportunism (think of the pirate ship). In these regards, as Sawday notes, Hamlet as engineer employs *machinations*—"a term which links machinery with deviousness or mental agility [and that] suggests how the term 'engineer' is cognate with the word 'ingenious'" (p. 99).[47]

---

[47] Sawday's reference to Hamlet's moon device falls within a chapter in which he is discussing the early modern publishing and printing phenomenon known as machine books (which were in fact sometimes called "theaters of machines")—works such as Vittorio Zonca's *Novo Teatro di machine et edificii* (Padua, 1607), Georg Andreas Boeckler's *Theatrum Machinarum Novum* (Nuremberg, 1662), and, most famously, Agostino Ramelli's *Le Diverse et Artificiose Machine* (Paris, 1588), among a great many others. These were "luxurious and collectable volumes":

Beautiful, costly, and sumptuous, in these works machines were disassembled, labelled, categorized, and explored. They were dedicated to noble patrons as advertisements for the mechanical ingenuity of their authors. Often plundering their designs from one another, within their pages the machine book authors laid out, in illustrative detail, the wonderful variety of mechanical creations which could be imagined (if not actually constructed) by the new generation of late sixteenth- and earlier seventeenth-century engineers. [p. 99]

III

In the final movement in this chapter I would like to consider another book that, like *Hamlet*, is interested in machines and the production of knowledge through the combination of naked-eye observation ("I'll observe his looks;/I'll tent him to the quick. If a do blench,/I'll know my course") and observational machines ("The play's the thing"). My purpose in turning to Tycho Brahe's 1598 *Astronomiae Instauratae Mechanica* is not to capitalize on an admittedly striking series of coincidences that, from a certain perspective, can be said to connect Tycho's book to *Hamlet* (or *Hamlet*—or Hamlet—to Tycho's book). Still less is my objective to establish a connection between these two works that is merely a matter of possible influence. Rather, my concern is first to work quickly through these preliminary matters and, second, to expand the notion of theater as machine to consider Tycho's efforts to recreate his observatory as itself a gigantic machine dedicated to astronomical observation and the production of celestial knowledge, especially in the form of quantified information (what we call "data"). And finally, I would like to speculate briefly—and to pose a number of questions and mark out a territory for further study—on the consequences (for astronomy, for science, and for the study of early modern culture more generally) of the process of what I want to call of *the textualization of the universe* that is nascent in Tycho and then more fully and more explicitly developed in Galileo's telescopic and mathematical astronomy, the subject of my next chapter.

The title page of *Astronomiae Instauratae Mechanica* (*Instruments for the Restoration of Astronomy*) features Tycho's likeness (at age 40, in 1586), set within an arch decorated with the names and coats-of-arms of his ancestors (Figure 3.1). Among the sixteen names that appear, two have caught the attention of a number of readers of Shakespeare's *Hamlet*: GVLDENSTEREN and ROSENKRANS. One of the earliest critical attempts to connect Shakespeare to this engraving of Tycho, and thereby to the two family names in question, was offered in 1938 by Leslie Hotson who suggested that Shakespeare may have been familiar with this image by virtue of an encounter with it in the house of Thomas Digges, the noted early modern English mathematician and astronomer and one of the earliest of the English advocates of Copernicus.[48] More recently, the astronomer and historian of science Owen Gingerich argued for a possible connection between Shakespeare and Tycho (and between Shakespeare and early modern astronomy more generally) based not solely upon the engraving, but also upon his reading of certain astrological references in the plays—and in *Hamlet* in particular—that serve to suggest that

---

[48] Hotson (1938: 123–4). For Hotson, this Shakespeare–Tycho connection (if it existed at all) was strictly a circumstantial matter of nomenclature, providing the dramatist with nothing more than likely Danish names. "Did [Shakespeare's] all-powerful imagination," Hotson wonders, "remain cold to the wild poetry of the astronomer's boundless enlargement of the Copernican universe?" He continues:

> Did it fail to discern in Heminges's scientist-neighbour the daring explorer of infinite space, who called the Earth 'this little dark star wherein we live'? One must admit that among Shakespeare's myriad minds, there was not the mind ready to kindle to the truth of Digges's vast vision. [pp. 122–3]

Figure 3.1 Author's portrait, Tycho Brahe, *Astronomiae Instaurate Mechanica* (1598)

Shakespeare had actually read Tycho's 1596 *Epistolae* (in which the same engraved portrait of Tycho had appeared).[49]

Those readers of the play who have cared to comment on this matter have taken up various positions along a spectrum marked, on the one hand, by a complete rejection of this idea, and, on the other hand, by something of a celebration of it and its consequences for reading *Hamlet*. As an instance of the former, Harold

---

[49] Gingerich (1981: 394–5).

Jenkins, the editor of the Arden 2 *Hamlet*, denies any necessary connection between these "splendidly resounding names" and Tycho, noting that these were perfectly common early modern Danish names, especially "among the most influential Danish families." Frederick II, Jenkins notes, "had nine Guildensterns and three Rosencrantzes at his court" and in the 1596 coronation procession for Frederick's successor, Christian IV, "one in ten bore one or other of these names." The "conjectures that Shakespeare knew of [the Tycho] engraving," Jenkins concludes, "are not necessary to account for a conjunction as natural as it is felicitous in giving an authentic touch of Denmark."[50] At the opposite extreme, the Tycho engraved portrait has inspired still another reader of *Hamlet* to offer an argument that connects the play not only to Digges and Tycho, but to Ptolemy and Copernicus: in a series of related articles, Peter Usher has constructed an elaborate—if unlikely—allegorical reading of the play according to which

> Prince Hamlet . . . personifies the New Astronomy which comprises the heliocentric model of Copernicus, and the Infinite Universe of stars of Thomas Digges. Claudius personifies the bounded geocentric model of his namesake Claudius Ptolemy, while Rosencrantz and Guildenstern personify the bounded model of the Danish astronomer Tycho Brahe (1546–1601). The allegory recounts the struggle to distinguish physical reality from appearances, a struggle well-known to astronomers as they strive to convert images [of] the sky to the reality of three- and four-dimensional space.[51]

To agree that "Rosencrantz" and "Guildenstern" together represent a point of connection between Shakespeare and Tycho Brahe is to reach a certain conclusion about Shakespeare and astronomy in general, and about *Hamlet* and Tycho in particular. To conclude, on the other hand, that these names were simply common enough to have come to Shakespeare's mind purely as names may have the effect of denying any explicit connection between Shakespeare and Tycho. While this question of possible influence may well be undecideable (though in the absence of any documentary evidence that links Shakespeare to Digges, I am inclined to believe the latter), the entire question is underwritten by another: was Shakespeare aware of, and therefore influenced by, contemporary astronomical study and research represented, for example, in Tycho's work?[52] Not only is this a familiar question, but it is a familiar *kind* of question about influence that holds to the conventional notion of a literary achievement set against a given cultural background—in this instance, the development of modern astronomy. As such, this question is of less value in a study such as the present one that is interested, in fact, in replacing such traditional—that is to say, unidirectional—models of influence (in a manner familiar since the advent of new historicist criticism, to cite one

---

[50] Shakespeare (1982: 422–3).
[51] Usher (2002: 144). This is the second of Usher's three articles to appear (perhaps oddly, given his general inattention to the so-called authorship issue) in *The Oxfordian*; see also Usher (2001), (2005), and (1999). Usher is a professor of astronomy and astrophysics—and webmaster of the Shakespeare Digges Homepage (<http://www.shakespearedigges.org/>).
[52] This is largely (though not wholly) a separate matter from the question of Shakespeare's relation to astrology.

precedent) with a more dynamic model of mutual construction and constitution across a given culture. There is, in other words, a more fruitful possibility: that there is a more important relation between Tycho and Hamlet's machine. Or, to phrase this differently: Tycho's astronomical work—especially the observational and computational work undertaken during his years at his Uraniborg observatory and represented in his book *Astronomiae Instauratae Mechanica*—is itself another manifestation of Hamlet's machine.

Tycho published *Astronomiae Instauratae Mechanica* in 1598, at the age of fifty-two and nearly a year after emigrating from his native Denmark.[53] Tycho's book is a detailed—and occasionally nostalgic—account (complete with illustrations and maps) of Uraniborg. The founding of Tycho's castle on the Danish island of Hven had been made possible by a grant—of land and revenue—from Frederick II, King of Denmark and Norway some twenty-one years earlier. Uraniborg was made famous by the careful and voluminous astronomical observations conducted there in what was effectively Europe's first purpose-built modern observatory.[54] Two of Uraniborg's defining features were especially significant: the organizational and institutional revolution that it constituted, on the one hand, and, on the other, its machinic nature.

Tycho's Uraniborg was revolutionary in its organization, its distribution of labor (both in its construction and in its observational endeavors), its fundamental investment in metrology, and—in a feature that helps make clear its status as an early instance of "Big Science"—the tremendous national expenditure required for its founding and continued operation. Frederick's grant of the Hven tenure, J. L Heilbron notes in *The Sun in the Church: Churches as solar observatories*, was reputed to have made Tycho "the richest scholar in Europe." Heilbron also remarks on the scale of the Uraniborg enterprise: "With assistants recruited mainly from Scandanavian universities, Tycho maintained a schedule of some 185 observing sessions a year and amassed a treasure of accurate, systematic data about the luminaries and the planets."[55] And in addition to this vast team of observers, and the assembly of observatories and observational instruments, Tycho's enterprise also included a working printing press and publishing office (as well as the construction of dedicated paper mills on the island), which was "the world's first working scientific press" and produced a number of pamphlets and books, including *De mundi aetherei recentioribus phaenomenis* [1588] and the *Epistolae astronomicae* [1596]).[56]

The second revolutionary feature of Tycho's enterprise follows as a consequence of the first: what I would like to call the *machinic* nature of Uraniborg itself. Tycho's Uraniborg did not merely house his astronomical instruments, but literally contained them within the very architectural fabric of its buildings.[57] There are a

---

[53] Brahe (1598). All quotations in the text are from Raeder, Strömgren, and Strömgren (1946).

[54] For discussions of Tycho's observatories, see—in addition to *Astronomiae Instauratae Mechanica*—Christianson (2000) and the most recent biography of Tycho, Thoren (with contributions by Christianson) (1990).

[55] Heilbron (1999: 10).

[56] Christianson (2000: 85), esp. Ch. 5 for a discussion of the story of Tycho's printing enterprise.

[57] This fact that lends special emphasis to the identification of Tycho offered in his engraved portrait as both the "*inventoris*" and "*structoris*" of the astronomical instruments of Uraniborg.

number of illustrations in *Astronomiae Instauratae Mechanica* that represent Tycho's Uraniborg as structurally an observational machine. One of these is the
revolving azimuth quadrant. In his description Tycho notes the great accuracy of
this instrument, which is able to determine altitude and azimuth with "the highest
accuracy and certainty, even within one-quarter of a minute of arc" (Tycho, 35).
This remarkable fact distinguishes this instrument from similar ones whose calculations cannot be so precise "partly on account of their small size, partly because they
are not built in such an ingenious way, nor so convenient to use." Any astronomer
who considers these facts, Tycho writes, "ought to consider this construction as
particularly commendable" (p. 35). In addition to the various steps surrounding the
instrument, which allow an observer to see the star in question either near the
horizon (from the highest step) or near its zenith (from the lowest), the great
precision this instrument affords the observer is a function of its entire structural—
one may say *architectural*—situatedness:

> This instrument with its small crypt-turret is covered on top by a roof, made of small,
> smooth beams, ingeniously joined together and connected, below the horizontal top of
> the wall and outside the azimuth circle, by a strong, round wooden ring. Hidden inside
> this ring are wheels, placed opposite each other in four places. With the aid of these
> wheels the roof can be turned around, with little effort, as may be desired. In this way
> the two oblong windows, which are placed in the roof opposite each other, and which
> are likewise formed of small beams, can be turned towards any star that is to be
> observed. [p. 35]

Tycho's description of his many instruments, which occupies the central portion of
his book and which is deeply invested in the architectural nature of their location,
placement, and uses, extends these same concerns in two further directions. The
first is what he calls Uraniborg's "architectonic structures suitable for astronomical
observations." Citing King Frederick's grant, Tycho writes:

> When thus according to the decision and gracious wish of this unique King I was to
> erect buildings which with their solidity and magnificent equipment were suitable for
> astronomical work, I chose the site highest on this island, which in itself is prominent,
> practically in its center, where I started building from its foundations a castle which was
> named after the heaven itself in which it was its task to make observations, and hence
> called Uraniborg. [p. 125–6] [58]

Tycho's second broadening gesture is to his consideration of the topography of the
island itself, in a section entitled "Topography of the Island of Venusia, popularly
called Hven, explanation of this topography":

---

[58] In a section of the Appendix entitled "Explanation of the design of Uraniborg with all its
premises," Tycho writes:

> You house dedicated to Urania, renowned beacon, erected in a high place and fortified with walls,
> surrounded by trees and lawns in your gardens, you, who in three times seven years have
> investigated all stars, while lifting your majestic head towards the Olymp—do you now stand
> unheeded? Do you stand silently and have been deserted? [1946: 127]

This island is situated in the far-famed Sound in the famous kingdom of Denmark, that which divides Scania from Sealand, and the capital Copenhagen is situated at a distance of three miles south-west, and Elsinore, with the Royal Custom House, is at a distance of two miles north north-west. [p. 138]

These passages suggest the degree to which the architectural and the topographical constructedness of Uraniborg are central to its functioning as an observational machine. As Elizabeth Spiller writes in *Science, Reading, and Renaissance Literature*, Tycho's description of Uraniborg "makes clear Tycho built on Hven what he understood to be an architectural expression of precisely the cosmic order that he hoped Uraniborg would enable him to discover, observe, and record."[59]

The illustration in *Astronomiae Instauratae Mechanica* that best represents the machinic nature of Tycho's entire enterprise is "The Mural, or Tychonian, Quadrant" (Figure 3.2). This is a complicated—because, in effect, a composite—image

Figure 3.2 "Quadrans Muralis Sive Tichonicus", Tycho Brahe, *Astronomiae Instaurate Mechanica* (1598)

[59] Spiller (2004: 126).

that is meant to represent not only the great brass wall quadrant itself (with a radius nearly two meters), but Tycho and the whole astronomical/scientific enterprise underway at Uraniborg. The engraving depicts the quadrant which was affixed to the wall of the room, together with the paintings that Tycho commissioned for the space created within the arc of the quadrant. The three paintings, by three different prominent artists of the day, depict the landscape vista (across the top of the frame); emblematical representations of Uraniborg within the arches of what the historian John Robert Christianson calls its "three regions": laboratory, museum, and observatory; and a formal portrait of Tycho himself (by the same artist who provided the model for the engraved portraits of Tycho discussed above) dominating the scene of the production of astronomical observation.[60] "The message of the painting," Christianson writes, "was that Tycho Brahe had created a research institute to probe the secrets of heaven and earth, and that the institute itself comprised a microcosm in harmony with the cosmos" (p. 118).

But if this is "the message of the painting," then the engraving that depicts the painting—together with figures that are not in the painting, including the quadrant, Tycho himself, and the various assistants measuring and recording astronomical details—has as its ambition the representation of representation itself, in its various forms, with its various objectives, including as perhaps the culminating act, the formalization and valorization of science that print affords.[61] At the same time—and even more significantly—the engraving represents the truly revolutionary ambition, and perhaps accomplishment, of the observatory-machine: the image radically displaces the universe that functions as the ostensible object of study. The celestial world to which the figure of the astronomer points does not so much lie beyond (or through) the quadrant's sighting aperture as *enter into* the interior space of Uraniborg where it can now be said to exist within the observatory-machine's various systems of representation: metrologically according the quadrant's values, symbolically through the painting's depiction of the work undertaken in Uraniborg (by globes and quadrants), archivally in the library and museum depicted, and textually in the register of observational data recorded in the assistant's book. By relocating God's celestial machine—what he calls the "wonderful clockwork of the heavens"[62]—within the very fabric of the observatory-machine, Tycho

---

[60] Christianson identifies the three painters: the landscape was painted by Hans Knieper, the views of Uraniborg by Hans van Steenwinckel, and the portrait by Tobias Gemperle.

[61] Questions concerning the relationship between early modern science and print culture constitute an important area of research within the current re-evaluation of science and culture. For recent contributions to this conversation, see, for example, Frasca-Spada and Jardine (2000) and Johns (2000).

[62] This phrase is itself a material feature of Tycho's observation-machine. In a section of the Appendix (entitled "Design of Stjernenborg, located outside Uraniborg"), Tycho transcribes an inscription "written in gold letters on a porphyry stone on the southern back of the portal":

Consecrated to the all-good, great God and Posterity . . . who will live for ever and ever, he, who has both begun and finished everything on this island, after erecting this monument, beseeches and adjures you that in honour of the eternal God, creator of the wonderful clockwork of the heavens, and for the propagation of the divine science and for the celebrity of the fatherland, you will constantly preserve it and not let it decay with old age or any other injury or be removed to any other place or in any way be molested, if for no other reason, at any rate out of reverence to the creator's eye, which watches over the universe. Greetings to you who read this and act accordingly. Farewell! [1946: 136]

substitutes one machine for another. In this way—by "incorpsing" the created universe within the created observational machine itself—Uraniborg both figures and embodies the universe it is dedicated to studying.

Tycho's success in embodying the visible universe literally within the fabric of his observational machine constitutes a critical moment in modern astronomy. In this regard, Uraniborg resides on the very cusp between pre-modern and modern science. On the one hand, Tycho's machine (in many ways) represents the culmination of naked-eye observational astronomy; on the other hand, if it is our first modern observatory (as I have suggested), it is also the *only* modern observatory without a telescope and in this sense, the last of a long line of non-modern observatories.[63] Within a decade of Tycho's death the very nature of astronomy—and, in a revolutionary way, the nature of the natural world—would be fundamentally changed by the introduction of the telescope. But what is it in particular that the telescope achieves and how are these achievements realized? What does the telescope do to the nature of astronomy—and to the nature of science? In the following chapter, these questions will lead to the figure of Galileo and his epochal introduction of the telescope—not only to astronomy, but to the popular imagination of the period, as well. In particular, Galileo's many innovations in astronomy, together with the resulting innovations in cosmology, are founded upon a revision of Tycho—not only in the commonly held notion that the Copernican model espoused (and partially established) by Galileo constitutes a final revision (or rejection) of Tycho's model, but also to the extent that Galileo's landmark observational and telescopic astronomical practice substantially revises Tycho's observatory-machine innovation.

Galileo achieves this revolution through the revision of two features of Tycho's Uraniborg astronomy. On the one hand, rather than conjuring a building without a telescope, Galileo offers instead a building that *is* a telescope. And on the other hand, Galileo revises Tycho's architectural and structural embodiment of the universe within the very fabric of his observatory, to the more radical reduction of the founding of the universe *within* the astronomical instrument par excellence: the refracting telescope itself.

---

[63] The great exception to this is, of course, Kepler, who is both a naked-eye astronomer and one of our first *modern* astronomers.

4

Galileo's Telescope

[G]l'attributi si devono accomodare all'essenza delle cose, e non
l'essenza à i nomi; perche prima furon le cose; e poi i nomi.

Galileo Galilei, *Istoria e dimostrazioni*
intorno alle macchie solari

I

In his 1613 book *Istoria e dimostrazioni intorno alle macchie solari e loro accidenti*
(*History and Demonstrations about Solar Spots and their Properties*, translated as
Letters on Sunspots), Galileo conjures a window. This window is both translucent,
rather than perfectly transparent—it is a stained glass window in a church—and
very slightly broken. These defining features allow sunlight to pass directly through
the space provided by its missing pane and in the process reveal to the careful
observer a remarkable effect: the production (through projection) of a striking
image:

> I have since been much impressed by the courtesy of nature, which thousands of years
> ago arranged a means by which we might come to notice these [sun] spots, and
> through them to discover things of greater consequence. For without any instruments,
> from any little hole through which sunlight passes, there emerges an image of the sun
> with its spots, and at a distance this becomes stamped upon any surface opposite the
> hole. It is true that these spots are not nearly as sharp as those seen through the
> telescope, but the majority of them may nevertheless be seen. If in church some day
> Your Excellency sees the light of the sun falling upon the pavement at a distance from
> some broken windowpane, you may catch this light upon a flat white sheet of paper,
> and there you will perceive the spots.[1]

In Galileo's account, the window is so important precisely for the gap or hole that it
frames (and therefore constitutes in a negative fashion) and which serves to let into
the interior of the church that single ray of sunlight. In this regard, the window
functions neither by virtue of letting in diffused light (a modified form of transpar-
ency that is itself a function of its nature as stained glass), nor because it is opaque
and therefore, in a manner of speaking, non-communicative and a mere thing.

---

[1] Galileo Galilei, *Letters on Sunspots*, in Drake (1957: 116–17). Unless otherwise noted, all
quotations are from this edition.

Galileo's church window functions instead because its strict disfunctionality (it is a *broken* window, after all) is of such a kind that renders the window instrumental within a larger device—which in a moment I will want to call a machine—that depends upon both the gap or hole in the window and the ability of the intact remainder of the window, in the manner of a wall, to hold out from the interior of the church any other rays of light.

But the first thing to point out, perhaps, in consideration of this scene within the church is that it is located within a fiction—even if that fiction posits a certain non-fictionality: "If in church some day . . . . " As such, Galileo's window (indeed, the church, the sunlight, the observers, and so forth) can only be said to be figural and not actual. The entire narrative, in fact, stands as a particularly interesting instance of one of Galileo's famous thought experiments. But its figural nature should not confuse us: it is a literal machine that Galileo imagines in his tale and therefore certainly not reducible simply to a trope. To be sure, Galileo's church-as-observatory does indeed carry a secondary meaning that is tropological in nature and centrally important to his efforts to avoid any further antagonism of the Roman Church. In the aftermath of Galileo's first astronomical book, the 1610 *Sidereus Nuncius* (to be taken up in more detail in the following section of this chapter) and the profound challenge it was thought—at least by some—to represent to Christian orthodoxy, Galileo was clearly interested in denying his antagonists any additional ammunition with which to continue, and even to escalate, their assaults. It is for this reason that Galileo's story of the Church is offered in the manner it is: as an attempt both to domesticate the celestial news his astronomical work was generating, and to make the case (if largely figurally) that the Church and the new astronomy not only could coexist, but indeed that they were mutually sustaining.

In his account, Galileo furthers this enterprise in rhetorically subtle fashion, presenting the solar effects his studies revealed—and their even more important consequences for cosmology more generally—not as the products of his efforts or imagination, but rather as simply matters of nature. As I will discuss below, Galileo was very keen to return to the debates about the validity of the telescope and this is perhaps the fundamental reason why he offers the story of the church-observatory not as an argument about the viability of his methodology, but rather as an illustration of the state of nature. The sun, Galileo declares, has, and has always had, spots on its surface; and these spots, moreover, have always been observable, even to the unaided eye: "For without any instruments, from any little hole through which sunlight passes, there emerges an image of the sun with its spots, and at a distance this becomes stamped upon any surface opposite" (*Letters*, 116–17).

This is powerful rhetoric, but even Galileo's rhetorical strengths as a writer, however prodigious they indeed were, cannot obscure completely the sleight-of-hand operating in this particular argument, since in the given example said to illustrate the natural and factual status of sunspots "without any instruments," the hole in the window, the surrounding walls, and the floor offered within the example can only be understood as a coordinating set of *instruments*. Windows, walls, and floors, for all of Galileo's masterful nonchalance here, are not naturally occurring objects. And even though this narrative about the nature of nature claims to illustrate that the image of the sun "becomes stamped upon any surface" exclusively through the agency of sunlight, this is manifestly not true. Or not the whole truth, since the "stamping" of that image can only be achieved through a certain instrumentality that serves to deny any strictly natural status to the phenomenon Galileo pretends merely to be reporting. It is especially important here to keep in mind that Galileo's account of the window—which I have called an account, a tale, a narrative, a story—is itself entirely imagined: it is a thought experiment rather than a strict retelling of some past event. Consequently, the apparently accidental— and therefore wholly natural—quality of the appearance of sunspots in the image projected through the broken window onto the floor of the church and that functions in Galileo's larger argument to establish nature as the ultimate object of his (scientific) inquiry, is a strictly *literary* effect: the scene is one that Galileo imagines and then conjures for his readers in order to point toward a truth that is larger than the simple story that he tells. In this regard Galileo can be said to offer something of a (scientific) parable in which the story told is only nominally about a church, but fundamentally about Galileo's efforts to communicate his understanding of the world made possible through his inquiry into nature and, equally, his interest in representing and communicating that greater understanding.

Galileo's window—and especially the gap through which the all-important ray of light passes—is only one part of a larger system that collectively works to produce a certain effect: the casting of the image of the sun onto a flat surface. This larger system is best understood, I want to say, as a machine. Like all machines, this one is an organized collection of instruments (tools) that work in collaboration to achieve a cumulative experimental—and experiential—effect that none could achieve separately, but which taken together they produce quite effectively. As such, Galileo's church-as-observatory satisfies the three characteristics identified earlier in this book as necessary to the machine. First, as "a stratagem, a kind of cunning," it represents a machination, in the model Bruno Latour has proposed.[2] Secondly, this machine—which can exist only by virtue of concerted and disciplined organization of human agency—certainly succeeds in the capture of nonhuman agency as figured in the real-time and dynamic visual effects projected by light in collaboration with the necessary opacity of the church onto its pavement floor. And thirdly, it produces a certain artificial experience: that of gazing directly onto the surface of the sun and perceiving there at least some of its more conspicuous visual material features.

In the story of Galileo's church-observatory, whatever spots may appear within the image of the sun "stamped" on the floor are (like the image itself) certainly projected through the machinations of the church-observatory and its silent— though certainly not natural—functioning as a machine. In other words, the sunspots Galileo projects in this thought experiment are, in every sense of the

---

[2] Latour (1987) writes, "A machine, as its name implies, is first of all, a machination, a stratagem, a kind of cunning, where borrowed forces keep one another in check so that none can fly apart" (p. 129). For a powerful consideration of the experiment and artificial experience, see Dear (2001).

term, produced by the orchestrated collaboration of the natural (sunlight, for instance) and the artificial (the absent individual pane framed by the rest of the window). In fact, this serves quite well as a description of what a machine is and what a machine achieves. Machines are never perfectly artificial; they always require a point of contact with the material world (nature) in order for them to function, in order for them to be machines. At the same time, and particularly within the scientific scenario illustrated in Galileo's parable and (as this chapter will argue more directly) in telescopic astronomy more generally, the natural cannot be said to exist wholly outside the mechanical, the technological, the artificial, the machinic.

In this light, then, Galileo's sunspots stand as apt illustrations of what Bruno Latour (following Michel Serres) has identified as "quasi-objects." In his important book *We Have Never Been Modern*, Latour extends "actor-network" (sociological) theory (pioneered by Latour, together with Michel Callon, and John Law, among others) toward science in general and the field of science studies in particular. Latour's notion of quasi-objects allows for our understanding of the "network" made up of humans together with non-humans that comes more clearly into view once we have divested ourselves of the notion (fully entrenched in the post-Kantian—that is to say, the modern—world) of a defining Nature–Society polarity. For Latour, the space along the axis that traditionally separates the Nature pole from the Society pole is not in fact empty, but rather full—of hybrids and quasi-objects, the very existence of which destabilizes not only the Nature–Society polarity, but equally both the purely realist logic, which would see truth located in nature, and the purely constructivist logic, which would see truth as the sole property of society. Latour writes:

> As soon as we grant historicity to all the actors so that we can accommodate the proliferation of quasi-objects, Nature and Society have no more existence than West and East. They become convenient and relative reference points that moderns use to differentiate intermediaries, some of which are called "natural" and others "social" . . . The analysts who head left will be called realists, while those who head right will be called constructivists.[3]

Quasi-objects reside not only between these two poles, but also "off" the horizontal axis that allegedly connects them; they are both "in between and below the two poles."[4]

The question remains, however, whether the same artifactual status of these sunspots holds true for those sunspots whose emergence into early seventeenth-

---

[3] Latour (1991: 85).
[4] Latour continues:

Quasi-objects are much more social, much more fabricated, much more collective than the "hard" parts of nature, but they are in no way the arbitrary receptacles of a full-fledged society. On the other hand they are much more real, nonhuman and objective than those shapeless screens on which society—for unknown reasons—needed to be "projected". By trying the impossible task of providing social explanations for hard scientific facts—after generations of social scientists had tried to denounce "soft" facts or to use hard sciences uncritically—science studies have forced everyone to rethink anew the role of objects in the construction of collectives, thus challenging philosophy. [Latour (1991: 55]

century consciousness follows principally from Galileo's work in both telescopic observation and print culture. It will be the argument of the rest of this chapter that this is precisely the case. Indeed, I will argue that the work achieved by Galileo's church-machine is a manifestation of his revolutionary use of the telescope. Using Galileo's astronomical work—beginning with his epochal *Sidereus Nuncius* and then continued to even greater effect in his work on sunspots (particularly as represented in *Letters on Sunspots*)—I will trace the emergence of the telescope itself as a scientific machine. The various features and facets of modern astronomical science established by Brahe in his Uraniborg-machine—observation, calculation, metrology, and printing—are in Galileo's hands and practice distilled into the telescope. Moreover, in a move that repeats Brahe's relocation of the "wonderful clockwork of the heavens," Galileo relocates the celestial machine a second time: where Brahe moved the machine of the universe into the observational machine of Uraniborg, Galileo moves the observational machine literally into the telescope itself.

At the same time, my concerns are also to consider the range of Galileo's astronomical work within the context of the emergence of science and to analyze the ways in which Galileo's first two major astronomical works—the 1610 *Sidereus Nuncius* and the 1613 *Letters on Sunspots*—mark respectively (and rather precisely) its two great phases. The first phase represents the discovery of things and the coming-into-being of objects and (nearly as often) the typically consequential passing-out-of-being of other objects. This sense of the first phase of scientific practice is in part underwritten by a desire (to borrow from Lorraine Daston) to describe "the historicity of scientific objects," the phrase Daston uses in her Introduction ("The coming into being of scientific objects") to a collection of essays titled, significantly, *Biographies of Scientific Objects*, to describe the dynamic processes whereby the objects of science can be understood to have "a different kind of reality than that set forth in the conventional two-valued metaphysics that obliges us to choose unequivocally between 'x exists'/'x does not exist' or 'x discovered'/'x is invented.'"[5] For Daston, the task (of her Introduction and the individual essays of the collection that it introduces) lies in the "attempt to revive ontology for historians" (p. 14).[6]

I will call this series of linked developments and events the *discovery* phase and I understand it to precede, as a matter of necessity, the following moment in which the status and nature of these newly secured objects are negotiated. This second phase I will call science proper and its defining maneuver is the transformation of objects into other objects. At the same time (and as will be discussed in greater detail below), this second phase carries with it a number of other important features, including a more explicit articulation and defense of methodology; the anticipatory or preemptive refutation of counter-argument; the inclusion and

---

[5] Daston (2001: 13). See also Rheinberger (1997).

[6] Daston continues, in part by way of a caution: "But history notoriously transforms all that it touches. An ontology that is true to objects that are at once real and historical has yet to come into being, but it is already clear that it will be an ontology in motion" (p. 14).

integration (in both method and outcome) of an actual machine characterized by its aptitude for observation, metrology, and inscription, and dedicated to fulfilling the demands of machination, the capture of nonhuman agency, and the production of artificial experience; and, finally, the introduction and elaborated use of full-scale scientific illustration, from diagrammatic to mathematical, and mimetic pictorial representation. Taken together, all of these features—which are fully deployed in *Letters on Sunspots*—constitute the work of what is recognizably science. To trace Galileo's movement through these two books is, then, to follow the emergence of science-as-machine.

II

Galileo's astronomical studies began with his epoch-making first celestial book, *Sidereus Nuncius* (translated variously as *The Sidereal Messenger*, *The Starry Messenger*, or *The Starry Message*) published in Venice in 1610.[7] It was this little book's "unfolding [of] great and wonderful sights" (as the title page reads) "about the face of the Moon, countless fixed stars, the Milky Way, nebulous stars, but especially about four planets flying around the star of Jupiter" that managed, in the words of Galileo's most recent English translator (and one of our best readers of Galileo's work) Albert van Helden, to set off "a chain of events that was to shake the intellectual edifice of Europe to its foundations" (*SN*, 1).[8] The unprecedented impact this book would make was the result, as Galileo understood, of two linked issues: the discoveries themselves together with their consequences for our understanding of the cosmos, on the one hand, and, on the other, the significance of the instrument that ushered in this new dawn of world knowledge, the telescope: "In this short treatise," Galileo writes at the opening of *Sidereus Nuncius*, "I propose great things for inspection and contemplation by every explorer of Nature. Great, I say, because of the excellence of the things themselves, because of their newness, unheard through the ages, and also because of the instrument with the benefit of which they makes themselves manifest to our sight" (*SN*, 35). Following Galileo's lead, I would like to address these two concerns in order: the significance and impact of these new objects, and the significance and impact of this new instrument.

Even as he will celebrate the revolutionary nature of *Sidereus Nuncius*, van Helden is also interested in identifying its curious status as a book of science. In

---

[7] Galileo (1989). For a brief discussion of Galileo's Latin title, see van Helden's Preface. All subsequent references are to this edition, abbreviated *SN*, followed by page number.

[8] Similar sentiments abound in the literature, including van Helden's comment in his Preface that "the discoveries announced by Galileo [in *Sidereus Nuncius*] changed the terms of the debate about the world systems." Van Helden continues:

> If they did not provide arguments to tip the balance in favor of the Copernican system, they were nevertheless decisive for they made the ancient authority—the foundation of the traditional system of natural philosophy—irrelevant. With *Sidereus Nuncius* we enter the modern world. [*SN*, vii–viii]

fact, van Helden's Preface to his edition of Galileo's landmark book begins by noting the curiously non-intellectual nature of the book:

> *Sidereus Nuncius* is not comparable to the great treatises that form the canon of the history of science. It had neither the staying power of Ptolemy's *Almagest* nor the synthetic power of Newton's *Principia*. In fact, as a scientific achievement it cannot be compared to Galileo's own later works, the *Dialogo* and the *Discorsi*. There is a good reason for this: *Sidereus Nuncius* was not so much a treatise as an announcement: in a few brief words, and in sober language, it told the learned community that a new age had begun and that the universe and the way in which it was studied would never be the same. (*SN*, vii)

The answer to this riddle, van Helden suggests, is not far to seek: although it was an "unprecedented sort of book" and although there cannot really be "doubt of Galileo's acute vision and brilliant mind," *Sidereus Nuncius*, he writes, "was not the product of an intellect but rather of an instrument!" (*SN*, vii). While I would like to defer until a later section of this chapter a discussion of the idea that *Sidereus Nuncius* was produced by the telescope—and assertion that (though meant here merely rhetorically) I aim to take absolutely seriously (and, moreover, certainly believe to be true)—it is important to this discussion to ponder the quality of *Sidereus Nuncius* literally as a message, or an announcement. The declamatory nature of Galileo's book is, in fact, a function of its virtually complete investment in the literary and pictorial representation of the work of discovery. For it is the careful staging of the discovery of things—and what I will call the production of objects— that constitutes the whole of *Sidereus Nuncius*. Galileo opens the book (after his remarkable dedication to Cosimo II, Grand Duke of Tuscany) by striking a tone of wonder and amazement at the marvels of nature revealed to us through the use of the telescope. "Certainly it is a great thing to add to the countless multitude of fixed stars," Galileo assures us. And: "It is most beautiful and pleasing to the eye to look upon the lunar body" through the telescope, and "it will be pleasing and most glorious to demonstrate" the true nature of nebulae and the Milky Way. "But what greatly exceeds all admiration" will be the description of "four wandering stars, known or observed by no one before us" (*SN*, 35–6).[9]

But the initial excitement of these discoveries (yet to be described) very quickly gives way to the details of their recitation. Galileo begins with a brief description of the telescope—his first hearing of the device (by way of a rumor) and its "wonderful effect" on sighting distant objects, and his progressive improvements of the device to the point at which he can declare having "constructed for myself an instrument so excellent that things seen through it appear about a thousand times larger and more than thirty times closer than when observed with the natural faculty only" (*SN*, 37–8). After this brief history (and short description of technical matters, such as a simple way to determine the magnification power of any telescope, as well as a description, complete with a rudimentary sketch, of the refraction of light rays

---

[9] For discussions of the role of wonder in early modern culture, and in early modern science, see Greenblatt (1991), Campbell (1999), and Daston and Park (1998).

achieved within the telescope through the combined effects of its shape and its lenses), Galileo turns his instrument and his attention to the moon and it is here that we witness the first of a series of the discovery of things. Galileo turns his telescope toward the moon and, with little concern to explain their natures, he begins to catalog the many things he sees through the telescope that are not visible to the unaided eye: countless small (and, he suggests correctly, newer) spots super-imposed upon the "darkish and rather large spots [that] are obvious to everyone," "depressions and bulges" on the surface, together with "chains of mountains and depths of valleys." In short and by "oft-repeated observations," Galileo writes, "we have been led to the conclusion that we certainly see the surface of the Moon to be not smooth, even, and perfectly spherical, as the great crowd of philosophers have believed about this and other heavenly bodies, but, on the contrary, to be uneven, rough, and crowded with depressions and bulges" (*SN*, 40).[10] Having thus drawn the moon and its new features, Galileo concludes his discussion of the moon with a landmark series of five illustrations, in the form of copper plates and a promise.[11]

The promise Galileo offers at the end of the lunar section of *Sidereus Nuncius* is to provide a more detailed elaboration in a future work of what is certainly the most significant consequence that follows from the earth–moon homology that his description of the new features of the moon suggests:

> We will say more in our *System of the World*, where with very many arguments and experiments a very strong reflection of solar light from the Earth is demonstrated to those who claim that the Earth is to be excluded from the dance of the stars, especially because she is devoid of motion and light. For we will demonstrate that she is movable and surpasses the Moon in brightness, and that she is not the dump heap of filth and dregs of the universe, and we will confirm this with innumerable arguments from nature. [*SN*, 57]

As suggested above, the work of science—the transformation of objects into other objects: here, the transformation of the earth and moon from opposites into homologues for one another—is postponed until that later project. Although he stops short of offering an explanation for this decision to defer to the later and as-yet unwritten book, it seems clear at the very least that Galileo simply did not yet have those "very many arguments and experiments" or those "innumerable arguments from nature" to offer. In other words, while he could represent, in both words and in graphic depictions, the emergence of the "great things" promised on the title page of *Sidereus Nuncius*, and while he could describe the many new objects in the universe (multitudes of new fixed stars, new spots on the moon, among them),

---

[10] Galileo's other important lunar discoveries include the uneven terminator that separate the lighted from the dark portions of the face of the moon; the uniformity of shape of surface craters as demonstrated by shadows; the rotation of the moon revealed by sunlight on mountains; and—perhaps most significantly—the phenomenon of "earthshine" (solar light reflected off the earth that illuminates the moon) that serves, in effect, to darken the moon and, at the same time, the darkness of all planets and satellites.

[11] These—and, indeed, all—the illustrations in *Sidereus Nuncius* have occasioned significant critical commentary; see, for instance, Edgerton (1984), Winkler and van Helden (1992), and Gingerich and van Helden (2003).

Galileo could not yet submit them to the scientific work of the second phase of the process sketched above. It was for this reason that Galileo could merely promise to do so in the future. For the moment, then, he could only offer descriptions of these new features of the universe. Speaking of the new fixed stars, for example, which emerge as new objects as a result of telescopic observation of the Milky Way, Galileo writes:

> What was observed by us . . . is the nature or matter of the Milky Way itself, which, with the aid of the spyglass, may be observed so well that all the disputes that for so many generations have vexed philosophers are destroyed by visible certainty, and we are liberated from wordy arguments. For the Galaxy is nothing else than a congeries of innumerable stars distributed in clusters. To whatever region of it you direct your spyglass, an immense number of stars immediately offer themselves to view, of which very many appear rather large and very conspicuous but the multitude of small ones is truly unfathomable. [*SN*, 62]

The final section of *Sidereus Nuncius* is devoted to the description of the Medicean stars (the four new moons of Jupiter). Galileo offers an interesting account of the emergence of these moons. With a mathematical precision, Galileo describes turning his telescope toward the constellations on the evening of January 7, 1610 "at the first hour of the night," spotting Jupiter and then noticing "that three little stars were positioned near him—small but very bright." Galileo continues:

> Although I believed them to be among the number of fixed stars, they nevertheless intrigued me because they appeared to be arranged exactly along a straight line and parallel to the ecliptic, and to be brighter than others of equal size. [*SN*, 64]

Galileo then immediately offers a series composed of dozens of (schematic) diagrams in illustration of "their disposition among themselves and with respect to Jupiter," including their appearance on the evening of 7 January, with two stars on the east and one on the west.

Galileo's discussion of his incremental realization of the nature of these stars as moons provides an enlightening narrative of the processes of discovery and the production of objects. Upon first seeing the curious arrangement of the three stars parallel to the plane of the ecliptic, they exist as nothing more than curiously arranged fixed stars of a special brightness. But on the next evening, the evening of 8 January, when "guided by I know not what fate," Galileo finds "a very different arrangement":

> For all three little stars were to the west of Jupiter and closer to each other than the previous night, and separated by equal intervals . . . Even though at this point I had by no means turned my thought to the mutual motions of these stars, yet I was aroused by the question of how Jupiter could be to the east of all the said fixed stars when the day before he had been to the west of two of them. [*SN*, 65]

Having withstood a brief period of doubt about the motion of Jupiter itself ("I was afraid," Galileo confesses, "that perhaps, contrary to the astronomical computations, his motion was direct and that, by his proper motion, he had bypassed those stars" [*SN*, 65]) and a perfectly cloudy night of 9 January, Galileo has the epiphanic moment:

> Then, on the tenth, the stars appeared in this position with regard to Jupiter. Only two stars were near him, both to the east. The third, as I thought, was hidden behind Jupiter. As before, they were in the same straight line with Jupiter and exactly aligned along the zodiac. When I saw this, and since I knew that such changes could in no way be assigned to Jupiter, and since I knew, moreover, that the observed stars were always the same ones (for no others, either preceding or following Jupiter, were present along the zodiac for a great distance), now, moving from doubt to astonishment, I found that the observed change was not in Jupiter but in the said stars. [*SN*, 65–6]

Sidereus Nuncius is so successful at chronicling the coming-into-being of objects discovered through Galileo's careful celestial telescopic observations that it was by itself almost enough to explode the Ptolemaic world system which, by comparison to the world on offer in Galileo's book, looks both confined in scope (nearly to the point of a claustrophobic narrowness), and altogether sparsely populated; where before *Sidereus Nuncius*, for example, Orion was entirely composed of nine stars, after Galileo's starry message, Orion's "original" nine stars are virtually lost among the more than 500 that are discovered there.

But for all of its spectacular power to expand the parameters and the objective population of the universe nearly to infinity, *Sidereus Nuncius*, as suggested above, provides very little analysis—very little science, I want to say—and therefore remains restricted largely to the discovery phase described above. One way in which it at least gestures toward the science that lies outside of itself, *Sidereus Nuncius* stands as a call to the work of science that will take over where discovery proper ends. Galileo explicitly calls upon other observers and astronomers to submit the new celestial objects he has discovered to careful metrological analysis: what are their relative sizes; how should the luminosity of the new fixed stars be gauged; and most importantly, what are the precise periods of the Medicean stars? For Galileo, analysis (and what I am calling science) depends very heavily upon the mathematical—or, more precisely, the *geometrical*—demonstration of the properties of his new objects.

So certain is Galileo in this conviction that he not only admits to it, but rather celebrates what he believes to be the fundamentally geometrical nature of nature. The famous passage appears in *Il Saggiatore* (*The Assayer*, 1623), Galileo's most explicit defense of the new astronomy and its methods, and deserves quoting at length:

> In Sarsi I seem to discern the firm belief that in philosophizing one must support oneself upon the opinion of some celebrated author, as if our minds ought to remain completely sterile and barren unless wedded to the reasoning of some other person. Possibly he things that philosophy is a book of fiction by some writer, like the *Iliad* or *Orlando Furioso*, productions in which the least important thing is whether what is written there is true. Well, Sarsi, that is not how matters stand. Philosophy is written in this grand book, the universe, which stands continually open to our gaze. But the book cannot be understood unless one first learns to comprehend the language and read the

letters in which it is composed. It is written in the language of mathematics, and its characters are triangles, circles, and other geometric figures without which it is humanly impossible to understand a single word of it; without these, one wanders about in a dark labyrinth.[12]

As this famous passage makes clear, for Galileo there is a fundamental difference between the mere existence of an object (a moon orbiting Jupiter, say) and its particular *meaning*. We can certainly gaze upon any particular material object for as long as we might like without necessarily moving any closer to understanding it because understanding requires that we "comprehend the language" in which the object (like a book) is written. It is to this logically and temporally second question that Galileo will turn his attention as he moves from discovery to science, from *Sidereus Nuncius* in 1610, his starry message about the new objects of the world, to *Letters on Sunspots* in 1613, the interpretation, as it were, of his earlier communication.

Galileo's labor in this task was made significantly more difficult to the extent that many of the claims he made in the book—claims that are essentially Copernican in nature—were manifestly counter-sensuous and counterintuitive. It is perhaps not too much of an exaggeration to say that regardless of how complicated it would become in later articulations (notably Ptolemy's), Aristotle's cosmology was grounded on two central and apparently irrefutable facts of everyday experience: unless artificially impeded, heavy objects by their very nature fall straight downward on a line toward the center of the earth; and the corollary axiom—equally foundational—was that unless artificially impeded, absolutely light objects (fire, for instance) move in a straight line in the opposite (upward) direction on a line toward the circumference of the dome of the sky. It was out of these two simple principles (motions) that Aristotle constructed a cosmology that would remain largely unchallenged for nearly 2,000 years.[13] Galileo rejects both premises, eliminating straight motion by demonstrating that all motion, terrestrial and celestial alike, is circular in nature, thereby relegating strictly linear motion the status of myth, and exploding the idea of an absolute and wholly unique center (as both the natural destination of heavy objects and, consequently, as the organizing focal point of the universe) with the introduction of multiple centers of circular celestial motion. With Galileo's discovery that like the earth, Jupiter is orbited by moons, the entire notion of the earth residing at the singular center of the universe could no longer hold.

But *Sidereus Nuncius* is strikingly devoid of even these kinds of generalizations and conclusions. Instead, the book works to catalog new-found objects and to offer rudimentary representations of them. Galileo remains content merely to demonstrate and celebrate the sheer existence of these "great things." While these discoveries *implied* several fundamental revisions of cosmological theory, the explicit articulation of these revisions, together with the deliberate articulation of the new

---

[12] *The Assayer*, in Drake (1957: 237–8).
[13] It was these unquestionable premises that Galileo dismantled in his work not only in astronomy, but in mechanics and physics more generally.

world they served to produce, find no place in *Sidereus Nuncius* and have to await subsequent books that Galileo would write, including *Letters on Sunspots* and, famously, his 1632 *Dialogo*. The analytical work of explaining those objects discovered through Galileo's telescope (what I want to call the *scientific* work that transforms objects into other kinds of objects) literally exceeds the declamatory nature of *Sidereus Nuncius*—which is to say, the nature of discovery—and must await the scientific work of *Letters on Sunspots* and other books.[14]

For Galileo, the particular method of this work was dependent upon the proper manufacture and use of the refracting telescope. The introduction of the telescope to scientific observation and investigation was indeed a revolution of tremendous significance since it not only represented the first time an instrument was used to enhance the human senses for the specific (epistemological) purpose of analytical observation, but also promised fundamental changes to cosmology and the standard view of the literal centrality of "man in the cosmos" that was the twin inheritance of 2,000 years of philosophy and orthodox Christian teaching.[15] Sensing both the momentous and potentially destabilizing nature of the advent of telescopic observation and innovation, Galileo was interested in underscoring the fact that sense perception enhanced through an instrument was still sense perception. Speaking in the opening paragraphs of *Sidereus Nuncius* about the pleasures afforded by telescopic observations of the moon, for example, "so that the diameter . . . appears as if it were thirty times, the surface nine-hundred times, and the solid body about twenty-seven thousand times larger than when observed only with the naked eye," Galileo asserts that features thus

---

[14] One of Galileo's earliest—and certainly his most important—supporters and advocates was the Imperial Mathematician to Rudolph II, Johannes Kepler, who responded virtually immediately to *Sidereus Nuncius* with his own publication in which he declares his complete faith in Galileo's discoveries, even without his own first-hand experience: "I may perhaps seem rash in accepting your claims so readily with no support from my own experience. But why should I not Believe a most learned mathematician, whose very style attests the soundness of his judgment?" (Kepler [1965: 12–3]). Kepler continues:

He has no intention of practicing deception in a bid for vulgar publicity, nor does he pretend to have seen what he has not seen. Because he loves the truth, he does not hesitate to oppose even the most familiar opinions, and to bear the jeers of the crowd with equanimity. Does he not make his writings public, and could he possibly hide any villainy that might be perpetrated? Shall I disparage him, a gentleman of Florence, for the things he has seen? Shall I with my poor vision disparage him with his keen sight? Shall he with his equipment of optical instruments be disparaged by me, who must use my naked eyes because I lack these aids? Shall I not have confidence in him, when he invites everybody to see the same sights, and what is of supreme importance, even offers his own instrument in order to gain support on the strength of observations? [p.13]

For an important analysis of Kepler's *Conversation* (as well as his *Dream* of 1634) and its relation to *Sidereus Nuncius*, see Spiller (2004: Ch. 3).

[15] The history of the introduction of the telescope to astronomical observation, as well as claims to priority in this use, are complicated stories (and largely outside the concerns of the present discussion). In addition to Galileo, there were a number of other early modern figures who seemed to have turned their own telescopes toward the night skies at nearly the same moment, among them Thomas Harriot in England who conducted both lunar and solar observations with his telescopes beginning in midsummer, 1609.

revealed were entirely observed through, rather than invented by, the object–
instrument–eye interaction:

> Anyone will then understand *with the certainty of the senses* that the Moon is by no
> means endowed with a smooth and polished surface, but is rough and uneven and, just
> as the face of the Earth itself, crowded everywhere with vast prominences, deep chasms,
> and convolutions. [*SN*, 35–6, emphasis added][16]

But, in spite of such assertions, it is transparently the case that as a work of
discovery and as a work of rhetoric, *Sidereus Nuncius* is thoroughly involved with
announcing and demonstrating both the coming-into-being of objects—telescopes,
lunar mountains and valleys, moons orbiting Jupiter, countless fixed stars—and the
passing-out-of-being of other objects: crystalline celestial spheres, epicycles, nebulae, and the Milky Way. To many readers, however, this process must have
seemed far less a matter of sense perception than such passages from the text
would argue. Indeed, when set against the prevailing Aristotelian cosmology that
also claimed to be based fundamentally on simple sense perception and logical
deduction, the claims of *Sidereus Nuncius*, we can well expect, would have been
thought more fancy than fact, especially initially.

The seriousness of the need to respond to Aristotle and the Peripatetics,
together with the need to overcome simple sense perception, became clear
immediately after the publication of *Sidereus Nuncius*, as a number of famous
anecdotes, as well as a number of infamous publications, attest. Less than two
months after the publication of *Sidereus Nuncius* (the book was published on
12 March 1610 in Venice) Galileo was in Bologna and there attempted to show
his newly discovered celestial objects to mathematicians, astronomers, and faculty
of the university, including Giovanni Antonio Magini. The attempt, however,
failed. This event was recorded in a number of letters written by Magini and one
of his students, Martin Horky. Horky's letter, written to Johannes Kepler (27
April 1610), although perhaps more vitriolic than most, is quite representative of
the defensive and rather aggressively dismissive remarks offered in response to
Galileo's discoveries:

> Galileo Galilei, the mathematician of Padua, came to us in Bologna and he brought
> with him that spyglass through which he sees four fictitious planets. On the twenty-
> fourth and twenty-fifth of April I never slept, day and night, but tested that instrument
> of Galileo's in innumerable ways, in these lower [earthly] as well as the higher [realms].
> On Earth it works miracles; in the heavens it deceives, for other fixed stars appear
> double . . . I have as witnesses most excellent men and most noble doctors, Antonio
> Roffeni, the most learned mathematician of the University of Bologna, and many

---

[16] A few pages later Galileo makes the philosophical consequences of this declaration of the identity
that obtains between the moon and the earth:

> By oft-repeated observations of [lunar craters] we have been led to the conclusion that we
> certainly see the surface of the Moon to be not smooth, even, and perfectly spherical, as the great
> crowd of philosophers have believed about this and other heavenly bodies, but, on the contrary,
> to be uneven, rough, and crowded with depressions and bulges. And it is like the face of the Earth
> itself, which is marked here and there with chains of mountains and depths of valleys. [*SN*, 40]

others, who with me in a house observed the heavens on the same night of 25 April, with Galileo himself present. But all acknowledged that the instrument deceived.[17]

Horky's experience of the failure of the telescope (if he is being honest in his letter) was not unique, as there were numerous other instances of men who agreed to test the instrument but who similarly experienced—and sometimes reported— its apparent failure. For his part, Horky would never relent in his criticisms of Galileo and his discoveries and, shortly after the Bologna incident, would go on to write one of the numerous books against Galileo and his telescope.[18] But there were also men who refused even to look through Galileo's telescope. One of these was Giulio Libri, professor of philosophy at Pisa and Padua, prompting the sharp-witted Galileo to say after Libri's death, that since he refused to look at these celestial objects while on earth, perhaps he would look at them on his way to heaven.[19]

In many ways the most serious objection to Galileo's discoveries was the argument suggesting that the telescope was fundamentally flawed as an observational tool because it produced false results. This possibility was addressed, if ironically and sarcastically, by Christoph Clavius, chief mathematician of the Collegio Romano. As reported to Galileo in a letter from his friend Lodovico Cardi da Cigoli (dated 1 October 1610) Clavius not only initially rejected Galileo's astronomical discoveries (though he was later to amend his position), but suggested that in order for the Medicean stars to exist it would be necessary for a telescope to manufacture them. Da Cigoli writes:

Questi Clavisi, che sono tutti, non credono nulla; et il Clavio fra gli altri, capo di tutti, disse a un mio amico, delle Quattro stelle, che se ne rideva, et che bisognierà fare uno ochiale che le faccia e poi le mostri, et che il Galileo tengha la sua oppinione et egli terrà la sua.

[None of the followers of Clavius believe anything [of Galileo's discoveries], and among the others, Clavius, who is their leader, said to a friend of mine that he laughs at the four [Medicean] stars and that it would be necessary to fashion a spyglass that produces them and then shows them, and that Galileo can keep his opinions and he will keep his own.] [*Opere Galileo*, 10: 442]

In his study of Clavius, *Between Copernicus and Galileo*, James M. Lattis discusses this letter and the dismissal by Clavius of Galileo's discoveries that it has been conventionally understood to communicate. Lattis argues that there is neither any evidence that this letter communicates anything more than hearsay, nor any

---

[17] Martin Horky to Johannes Kepler, quoted in Galileo (1989: 92–3). Horky seems to revel in this narrative:

Galileo became silent, and on the twenty-sixth, a Monday, dejected, he took his leave from Mr. Magini very early in the morning. And he gave no thanks for the favors and the many thoughts, because, full of himself, he hawked a fable....Thus the wretched Galileo left Bologna with his spyglass on the twenty-sixth. (*SN*, 93)

[18] See Horky (1610).
[19] Galileo (1890–1909: X: 484), quoted in Drake (1957: 73).

evidence in Clavius's behavior—toward Galileo or as concerns the Collegio's own telescopic observations during the summer of 1610—to support the interpretation that Drake (for example) reaches that "Clavius had . . . been reported as saying that in order to see such things one would first have to put them inside the telescope" (*Letters*, 75). "[I]t would certainly be going too far," Lattis writes, to endorse Drake's reading:

> Not only does Drake here trivialize what is a nontrivial optical problem, but his harsh interpretation suggests that Clavius imputed intent of fraud to Galileo, which is not justified by Cigoli's own words.[20]

The suggestion that the objects Galileo saw through his telescope were illusions generated by the instrument itself was a common reaction to the potentially destabilizing fact of their actual celestial existence. Naturally, Galileo rejected all of these accusations, having submitted his findings to laborious repetition. And Galileo's results were, in quite short order, corroborated by astronomers and other authorities, including Clavius and the group of Jesuit astronomers and mathematicians appointed by Cardinal Bellarmine to pass judgment on Galileo's astronomical findings. But as I will argue in the following section, in spite of Lattis's more literally accurate translation of the Cigoli letter, Drake was nearer the mark in characterizing Clavius's sentiment as suggesting that in order to see Galileo's astronomical objects "one would first have to put them inside the telescope." And Drake was more correct, I would argue, because Clavius was more correct in his initial assessment of the telescope's *manufacture* of celestial objects than either he or Galileo imagined. In fact, invoking my earlier discussion of the telescope as machine and the status of sunspots as quasi-objects, I want to suggest that Clavius is exactly correct when he asserts that the objects Galileo discovered were produced by the telescope—in collaboration with the material and physical world—and were, as Clavius further mused, *shown* by it to the witnessing world. In order to establish this more fully, I will need to turn to Galileo's *Letters on Sunspots*, the text in which Galileo not only effected the move from the discovery phase of science (as I argued above was illustrated in his *Sidereus Nuncius*) to the analytical phase of science proper, but also in which the artifactual status of his newly discovered celestial objects is made clear.

For Galileo, the road to the definitive establishment of the validity of telescopic observation as bona fide sense perception led through ancient philosophy and the figure of Aristotle. Between the publication of *Sidereus Nuncius* in 1610 and *Letters*

---

[20] Lattis (1984: 184). For his part, Lattis translates the critical passage of Cigoli's letter thus: "Clavius, among others, and the head of them all, said to a friend of mine concerning the four [Medicean] stars that [Clavius] was laughing about them and that one would first have to build a spyglass that creates them, and [only] then would it show them. And [Clavius] said that Galileo should keep his own opinion and he [Clavius] would keep his" (p. 184). For another discussion of this letter, see Galileo (1989: 109). In addition to his work as an architect, Cigoli was also a significant painter, in which capacity he painted the Pauline Chapel of Rome's Santa Maria Maggiore in 1611 with a painting entitled *The Woman of the Apocalypse* depicting the Virgin standing upon a moon that is both spotted and mountainous. For a discussion of Cigoli's painting in relation to *Sidereus Nuncius*, see Booth and van Helden (2001). See also, Reeves (1997).

on Sunspots in 1613, Galileo was at work on a number of projects, some of which were astronomical (largely on the level of observation) and some not; one of these latter projects was his study of hydrostatics that led ultimately to the publication of his *Discourse on Bodies in Water*, published in early 1612.[21] Not only does this work of physics appear in print between the publication of two of Galileo's landmark astronomical books, but to the considerable extent that it addresses itself to Galileo's understanding of Aristotle and Galileo's divergence from Aristotelian (and Peripatetic) philosophy, the *Discourse on Bodies in Water* functions as the text—and also as the argument—that allows Galileo to make his philosophical way from the declamatory discourse of discovery represented in 1610 to the scientific discourse of 1613.

III

Galileo's interest in writing on hydrostatics was in part inspired by a controversial debate in Florence and Pisa (conducted by a number of prominent figures, including Filippo Salviati and several university mathematicians) over the nature of bodies in water and whether or not shape played any role in a given object either floating or sinking. In lieu of an actual public debate (which was to include demonstrations), Galileo wrote an essay, addressed to Cosmio II, the Grand Duke of Tuscany, that both sketched the history of the controversy and provided a brief and preliminary overview of Galileo's thoughts on floating bodies.[22] In this essay—which was later superseded by the vastly expanded *Discourse*—Galileo offers one of his most explicit formulations of the fundamental distinction between his work and that of Aristotle and the Peripatetics: when confronted with the "malignity, envy, or ignorance" of his opponents, Galileo assures his patron, he will "always put down (and with very little trouble) their every impudence," protected, as he says, by "the invincible shield of truth." Galileo continues, drawing his attention to the question of Aristotle:

> what I have asserted in the past was and is absolutely true, and that to the extent that I have departed from the commonly accepted Peripatetic opinions, this has come about not from my not having read Aristotle or not having understood as well as they his reasoning, but because I possess stronger demonstrations and more evident experiments than theirs.[23]

---

[21] Unless otherwise noted, quotations of the *Discourse on Bodies is Water* (abbreviated here as *Bodies*) are from Stillman Drake's edition of Thomas Salusbury's 1663 English translation titled *A Discourse Presented to the Most Serene Don Cosimo II. Great Duke of Tuscany, concerning the Natation of Bodies Upon, and Submersion in the Water. By Galileus Galilei: Philosopher and Mathematician unto His most Serene Highnesse*, Drake (1960).
[22] Drake provides a useful summary of the development of this debate, and of Galileo's essay and the writing of the *Discourse*; see Drake (1978, esp. 169–79).
[23] Quoted in Drake (1978: 172), hereafter *GAW*.

Galileo continues this powerful defense of his methods (to say nothing of his ethics and integrity, always also targeted by his rivals) by pointing precisely at what lies at the foundation of what he takes to be his better claims to scientific certainty, namely his greater attention to the investigation of causes:

> And in the present dispute, in addition to showing the approach I take to the study of Aristotle, I shall reveal whether I have well understood his meaning in two or three readings only, compared with them, to some of whom the reading of Aristotle fifty times may seem a small matter; and then shall I show whether I have perhaps better investigated the causes of the matters which constitute the subject of the present contest than did Aristotle . . . [24]

His essay continues, casting a glance back at the experimental work on floating bodies that he has accomplished (and that has, he argues, rendered Aristotle's arguments on the matter permanently obsolete), while at the same time registering his uncertainty at the response of the Peripatetics:

> I know not whether the adversaries will give me credit for the work thus accomplished, or whether they, finding themselves under a strict oath obliged to sustain religiously every decree of Aristotle (perhaps fearing that if disdained he might invoke to their destruction a great company of his most invincible heroes), have resolved to choke me off and exterminate me as a profaner of his sacred laws. [25]

In the opening moments of *Bodies in Water* itself, Galileo again addresses the matter of Aristotle and the tendency of the Peripatetics to consider his "decrees" effectively as "sacred laws" and any argument contrary to his philosophical texts as a kind of sacrilege. Galileo's response is not to deny that his work "is different from that of *Aristotle*; and interferes with his Principles," but rather to embrace these differences and, moreover, to claim Aristotle himself as the precedent for doing so:

> I hope to demonstrate that it was not out of capritiousnesse, or for that I had not read or understood *Aristotle*, that I sometimes swerve from his opinion, but because severall Reasons perswade me to it, and the same *Aristotle* hath tought me to fix my judgment on that which is grounded upon Reason, and not on the bare Authority of the Master. [*Bodies*, 3]

And again later in the text, when responding to the critique of Archimedes offered by Galileo's contemporary, the Peripatetic philosopher Francesco Buonamici, who faults Archimedes for what Galileo calls "some Repugnances to the Doctrine and

<hr/>

[24] *GAW*, 172. As this passage also demonstrates, Galileo's method is not only a matter of investigation (which, if broken into its constituent parts, consists of observation, hypothesis, and experiment), but also a matter of texts and reading. It is an important feature of Galileo's science that he should establish his credentials as a reader of Aristotle, *especially* since he intends to move beyond Aristotle.
[25] Galileo offers here an interesting literary analogue:

> In this they would imitate the inhabitants of the Isle of Pianto when, angered against Orlando, in recompense for his having liberated so many innocent virgins from the horrible holocaust of the monster, they moved against him, lamenting their strange religion and vainly fearing the wrath of Proteus, terrified of submersion in the vast ocean. [*GAW*, 173–4]

Opinion of *Aristotle*," Galileo addresses the divide between mere authority and reason:

> In answer to which Objections, I say, first, That the being of *Archimedes* Doctrine, simply different from the Doctrine of *Aristotle*, ought not to move any to suspect it, there being no cause, why the Authority of this should be preferred to the Authority of the other: but, because, where the decrees of Nature are indifferently exposed to the intellectuall eyes of each, the Authority of the one and the other, loseth all a[u] thenticalness of Perswasion, the absolute power residing in Reason . . . [*Bodies*, 20][26]

This notion of authority versus reason is something of a constant theme—not only in Galileo's book on floating bodies, but in virtually all of his work, both in science (in mechanics, telescopic astronomy, physics) and in his *defenses* of these scientific works, and, most significantly, in his trial by the Office of the Inquisition.[27] But in the study of floating bodies—and in 1611, between his two great astronomical texts—Galileo for the first time offers a sustained critique of inherited authority and does so, moreover, by an insistence upon demonstration and experiment.

What I am characterizing as Galileo's turn toward intensive observation, demonstration, and experimentation was certainly in process before he turned his attention to the problems of hydrostatics, and, if anything, accelerates in the years following the publication on floating bodies. But it is in this work that Galileo allows himself to fashion his overall argument as an example of reason triumphant over authority. In his introduction to Salusbury's translation of the *Discourse on Bodies in Water*, Stillman Drake points directly at the definitive nature of Galileo's critique of Aristotelian philosophy. In shifting from his essay to his book, Drake writes, Galileo "did not hesitate to turn his book into a bold and uncompromising blow against the very foundation of Aristotelian physics." Drake offers two reasons for the book's great effectiveness. The first is Galileo's decision to publish in "the colloquial Italian" and while this is not wholly unprecedented, it was nearly so (most printed intellectual work in Europe typically appeared in Latin)—and marked both a new direction for Galileo (the starry message was in Latin), and at the same time began the series of vernacular publications that would last until the end of Galileo's career. Drake's other observation points to Galileo's deep investment in experiment:

> Perhaps of equal importance [to publication in Italian] was the simplicity and inherent interest of the experiments described, the relevance of which to the decision of the controversy was indisputable. This could not fail to weaken the long-standing tradition under which such disputes were customarily settled by appeal to authority, a tradition that remained the chief obstacle to the development of modern scientific notions. [*Bodies*, xxi]

---

[26] Buonamici was the author of the highly influential Aristotelian study of motion, *De Motu* (Florence, 1591); Drake suggests that Galileo "was almost certainly one of his pupils" (*GAW*, 442).
[27] Drake (1957) prints two of these important texts: *Letter to the Grand Duchess Christina* (1615) and (in abbreviated form) *The Assayer* (1623).

The status of *Discourse on Bodies in Water* as something of a necessary bridge that links Galileo's two great astronomical studies—and at the same time that it provides the philosophical legitimation of his scientific practices in the de-legitimation of Aristotelian and Peripatetic philosophy—is also signaled in the appearance in the opening pages of the book of Galileo's most recent astronomical studies. These include not only direct and immediate reference to the discoveries of *Sidereus Nuncius*—the true appearance of the surface of the moon, the "tricorporeall" shape of Saturn, the phases of Venus, and the "Four Medicean Planets about Jupiter"— but, significantly, Galileo's further refinements of his knowledge of the periods of the Jovian moons "which I lighted upon in April the year past, 1611, at my being in Rome" (*Bodies*, 1). In addition to the already published discoveries of *Sidereus Nuncius* and the more exact calculation of the correct periods of Jupiter's moons, Galileo also declares that he will continue his observations in the service of calculating the relative sizes of the moons. And, having "hit upon a way of taking such measures without failing," Galileo assures his readers that his methods "shall serve to bring us to the perfect knowledge of the Motions, and Magnitudes of the Orbes of the said Planets, together with some other consequences thence arising" (*Bodies*, 2).

But the most important announcement concerning astronomical observation and the ensuing consequences of their outcomes is reserved for something wholly new. Nothing in *Sidereus Nuncius* would have prepared Galileo's readers for the appearance in his *Discourse on Bodies in Water* for the mention of sunspots. There is little fanfare here—and not a little uncertainty, as well:

> I adde to these things the observation of some obscure Spots, which are discovered in the Solar Body, which changing, position in that, propounds to our consideration a great argument either that the Sun revolves in it selfe, or that perhaps other Starrs, in like manner as *Venus* and *Mercury*, revolve about it, invisible in other times, by reason of their small digressions, lesse than that of Mercury, and only visible when they interpose between the Sun and our eye, or else hint the truth of both this and that; the certainty of which things ought not to be contemned, nor omitted. [*Bodies*, 2]

So reads the first edition of *Discourse*. This edition sold out so rapidly that Galileo issued a second within a few months.[28] By the time of the second printing, Galileo was able to provide even more current and updated information on his observation of sunspots and his speculations on their natures:

> Continuall observation hath at last assured me that these Spots are matters contiguous to the Body of the Sun, there continually produced in great number, and afterwards dissolved, some in a shorter, some in a longer time, and to be by the Conversion or Revolution of the Sun in it selfe, which in a Lunar Moneth, or thereabouts, finisheth

---

[28] Drake points directly to its unique status of the second edition of the *Discourse of Bodies in Water* among Galileo's works for the attention it received from the author: "At its time it was widely read; indeed, the first edition was so speedily exhausted that Galileo prepared a second, with additions, which was published the same year. It is the only one of his books that received such treatment at his hands, though most of them were quickly sold out" (*Bodies*, ix).

its Period, caried about in a Circle, an accident great of it selfe, and greater for its Consequences. [*Bodies*, 2][29]

As Galileo's subsequent astronomical work would make increasingly clear, the discovery of sunspots *as objects of observational and experimental knowledge* helped to dismantle Aristotelian and Ptolemaic cosmology altogether. Indeed, one of the great revolutions in early modern cosmology was occasioned by the discovery of these blemishes on what was long believed to be the pristine (or the "immaculate") surface of the sun which could in no way be accommodated to the standard model of the universe. As he will go on to demonstrate in his book on sunspots, Galileo understood this discovery of dark spots on the sun to constitute definitive material proof against the presiding Aristotelian belief in a materially perfect universe. "[The sun's] supposed immaculacy," writes Galileo, "must yield to observation."[30]

Galileo wrote *Letters on Sunspots* in part as a refutation of a slightly earlier series of pronouncements by the German Jesuit writer Christopher Scheiner—cast in the form of a series of letters to Mark Wesler (an accomplished amateur scientist) who not only published Scheiner's letters but also sent a copy to Galileo for his evaluation.[31] At the same time, *Letters on Sunspots* constitutes the results of Galileo's own attempts—underway even prior to Scheiner's pamphlets—to solve the mysteries that the spots represented: were these spots simply minute flaws in the lenses of the telescope and therefore wholly illusory? Were they manifestations of refraction that occurred only in the action of visual perception and hence literally in the eye of the beholder? Were they instead previously undiscovered planets, as Scheiner proposed? Or did they actually exist on the surface of the sun. And if they did indeed exist, what then were they and what consequences followed from their sudden emergence into both the scientific and the popular imagination?

Part of the challenge Galileo assumed in his work on sunspots was the definitive demonstration of certain characteristics or properties of the sunspots: that unlike planets, the moon, and the sun, they were irregular in both appearance and motion; that they were migratory and transient; and that they were not "beneath the sun" but indeed on (or virtually on) the surface of the sun itself.[32] The first step, then, of

---

[29] These constitute, as Stillman Drake has discussed, Galileo's first published remarks on sunspots to appear in print, though later than the publication of a work on sunspots by the German Jesuit Christoph Scheiner. "Early in 1612," Drake writes, "after recovering from a long illness and while the *Discourse* was in the hands of the printer, Galileo conducted a series of careful observations which enabled him to refute utterly the contentions of Scheiner" (*Bodies*, 82).

[30] Galileo, *Letters on Sunspots*, in Drake (1957: 135).

[31] For a discussion of Scheiner and Wesler, as well as the publication of Scheiner's *Tres Epistolae de Maculis Solaribus Scriptae ad Marcum Welsrum* (1612), see Drake (1957: 81–5) and *GAW*, Chs. 9–11, and *passim*.

[32] Near the beginning of his third letter to Wesler, Galileo offers a moving discussion about the limits of human knowledge and understanding. In our speculating, Galileo writes, "we either seek to penetrate the true and internal essences of natural substances, or content ourselves with knowledge of some of their properties. The former I hold to be as impossible an undertaking with regard to the closest elemental substances as with more remote celestial things." He then offers as an example the inquiry into the substance of clouds: when told that clouds are "moist vapor I shall wish to know in turn what vapor is." And as from clouds to vapor, so too from vapor to water "attenuated by heat," and from water to "this fluid body which runs in our rivers and which we constantly handle." But even,

this work was establishing a certain thing-ness for sunspots. As Galileo and other early students of the spots knew perfectly well, even the task of establishing the sheer physical existence was a major challenge in the struggle to understand them. Being a mere thing, it turns out, is no easy thing. Once granted status as things, sunspots virtually instantly became objects of speculation, of debate, and most decisively of study. But this transformation from nothing (merely blemishes in the telescope lenses, as some argued) to object was not sufficient to determine the true nature of sunspots and earn them a definitively known existence, what we might want to call an ontology of their own. The next phase of the evolution of sunspots toward total existence, during which the arguments and debates continued, was the most contested and most interesting because it was during this phase that we can witness not the coming-into-being of objects (as in the first phase), but rather the transformation of objects into *other* objects.

Galileo makes his conclusions on the debates over the very nature of sunspots entirely clear early in his book: "I have no doubt whatever that they are real objects and not mere appearances or illusions of the eye or of the lenses of the telescope" (*Letters*, 90–1). They are as real, he insists, as are the dark spots on the moon which are produced "by projected shadows of lunar mountains" (another Galilean discovery) which "in comparison with the lighted portions they are as dark as is the ink with respect to this paper" (*Letters*, 93). This assertion enacts an interesting and important series of rhetorical and evidential substitutions: Galileo seeks to explain the appearance of sunspots by establishing their analogical relation to lunar shadows that are produced by the lunar mountains his own telescope had discovered (and which he had sought to represent through the series of illustrations in *Sidereus Nuncius*). This stands as the first analogy. The second is equally important and even more striking: both areas of celestial darkness—spots on the surface of the sun and the shadows of mountains cast on the surface of the moon—are analogous to the absolute darkness of ink *stamped* onto white paper.[33]

On the one hand, there is an unmistakable bravado in this gesture. As Stillman Drake has pointed out, of all the discoveries announced in *Sidereus Nuncius*, it was Galileo's claims of a mountainous lunar surface that were most unsettling to the cosmological status quo: "Any challenge to perfect sphericity of the perfect heavenly

---

Galileo writes, "this final information about water" is "no more intimate than what I knew about clouds in the first place; it is merely closer at hand and dependent upon more of the senses." And he concludes:

> In the same way I know no more about the true essences of earth or fire than about those of the moor or sun, for that knowledge is withheld from us, and is not to be understood until we reach the state of blessedness.

But the end is not despair:

> But if what we wish to fix in our minds is the apprehension of some properties of things, then it seems to me that we need not despair of our ability to acquire this respecting distant bodies just as well as those close at hand—and perhaps in some cases even more precisely in the former than the latter. [*Letters*, 123–4]

[33] Significantly, the term Galileo uses to characterize the appearance of sunspots on the white paper—or, the metaphor he uses to describe the action of sunlight—is "stamping" (*stampata*) in the model (I would argue) of printing (Galileo [1890–1909: V: 137]).

bodies constituted the worst threat of all in the eyes of philosophers, whose reasons in support of this dogma (despite contrary telescopic evidence) outnumbered their arguments against any other revelation made in the *Starry Messenger*" (*GAW*, 184).[34] Drake also quotes, at some length, the letter Galileo wrote to Federico Cesi that accompanied his first sunspot letter to Wesler in which Galileo states both his definitive conclusions about sunspots and the nature of solar motion, and the demise of the Aristotelian dogmas of the Peripatetics:

> I have finally concluded, and believe I can demonstrate necessarily, that [sunspots] are contiguous to the surface of the solar body, where they are continually generated and dissolved, just like clouds around the earth, and are carried around by the sun itself, which turns on itself in a lunar month with a revolution similar [in direction] to those of the planets . . . which news I think will be the funeral, or rather the extremity and Last Judgment of pseudophilosophy, of which [event] signs were already seen in the stars, in the moon, and in the sun. I wait to hear spoutings of great things from the Peripetate to maintain the immutability of the skies, which I don't know how can be saved and covered up, since the sun itself indicated [mutations] to us with most manifest sensible experiences. [*GAW*, 183][35]

In the face of this news, Galileo suggests, even the lunar mountains controversy will appear insignificant: "Hence I expect that the mountains of the moon will be converted into a joke and a pleasantry in comparison with the whips of these clouds, vapors, and smokings that are being continually produced, moved, and dissolved on the face of the sun" (*GAW*, 183).

On the other hand, the analogy to ink and paper—to print, I want to say—is perhaps even more revealing and suggests an alternative way to understand Galileo's sunspots. For although Galileo was correct in asserting that sunspots are absolutely bright and dark only relative to the rest of the sun (and though he was also right to say that we need not "assume the material of the sunspots to be very dense and opaque, as we may reasonably suppose with regard to the material of the moon and the planets" [*Letters*, 93]), I want to argue that whatever else they would become in Galileo's hands—proof of the sphericity of the sun; proof of the sun's rotation upon its axis; and, finally, proof of the sun's stationary position within the solar system and the earth's motion around it (an argument begun in *Letters* but not fully demonstrated until his 1632 *Dialogue Concerning the Two Chief World Systems*)— Galileo's sunspots emerge *as objects of scientific knowledge* strictly as a consequence of their machinic production. And as such, they not only reveal their existence to be artifactual, but more specifically *textual* in nature. If we recall Galileo's church-observatory parable discussed at the beginning of this chapter and consider again the argument there about the hybrid nature of sunspots projected onto the floor of the church—their status, that is, as simultaneously natural and artificial—then we can say in the instance of *Letters on Sunspots* that Galileo does not so much discover

---

[34] Clavius "had made this his one reservation about Galileo's announced discoveries" (*GAW*, 183).
[35] Federico Cesi was the founder of the Lincean Academy, to which Galileo was elected member in 1611; the Lincei undertook the publication of Galileo's *Letters on Sunspots*. For an important discussion of Cesi and the Lincei, see Freedberg (2002).

sunspots as help *to produce* them.[36] They are quasi-objects revealed by the operations of an actual machine composed of many parts, including both natural components (light, most significantly), and manufactured components: the lenses of the telescope. And if we consider further Galileo's innovative use of the telescope not only as an observational tool, but even more importantly *as a writing instrument*, then we can reasonably conclude that any existence that could be claimed for the sunspots was indeed a matter of "ink with respect to . . . paper."

Galileo can construct a more complete and accurate theory of sunspots (and the cosmological consequences that follow) in large part because he has mastered the machinery of a new method of representation. It is Galileo's machinic and textual practices that not only make his arguments stronger than Scheiner's, but that also allows Galileo the very means through which he can produce the science of sunspots, something Scheiner was not able to do until a number of years later and only once he abandoned his methods and adopted Galileo's wholesale.[37] In other words, Galileo's sunspots do not simply emerge from the surface of the sun into human understanding as the result of the observation of nature, but rather from the very technology that enables such observation in the first place. Sunspots, we can say, were invented by the collaboration of nature and artifice under the direction of Galileo who manipulates both the telescope and the natural light it captures and employs both in the service of a technology of representation devised in order to make the sunspots manifest both to the scientist and subsequently—through the technologies of illustration and print—to his readers.

Scheiner's observational and representational methods were, in the strict sense, rudimentary in nature (Figure 4.1). Scheiner covered the eyepiece of the telescope with colored glass, peered through the telescope at the sun, then drew freehand what he saw onto a piece of paper. This method, though it has the virtues of simplicity, also has a number of significant liabilities. It required the frequent shift of the gaze from the lens of the telescope to the drawing paper and then back again; by its very impressionistic nature it eliminated the possibility of accuracy of both location and placement; it similarly disallowed drawing to scale; it made possible the observation and representation of only the largest of the sunspots; and it provided no structural guarantee against distortion.

For his part, Galileo is careful to offer an elaborate discussion of his method, which is based upon the innovations established by one of his students, Benedetto Castelli, "of a noble family of Brescia—a man," Galileo writes, "of excellent mind, and free (as one must be) in philosophizing."[38] The description of this method that Galileo offers is worth quoting at some length:

---

[36] These terms—"natural" and "artificial"—are related to the terms "fact" and "artifact" as deployed in the foundational book by Latour and Woolgar (1979); see especially Ch. 4, "The microprocessing of facts."

[37] See Scheiner (1630).

[38] Drake identifies Castelli as both "the greatest of Galileo's scientific pupils" and an important teacher in his own right whose own students included Evangelista Torricelli (1957: 115).

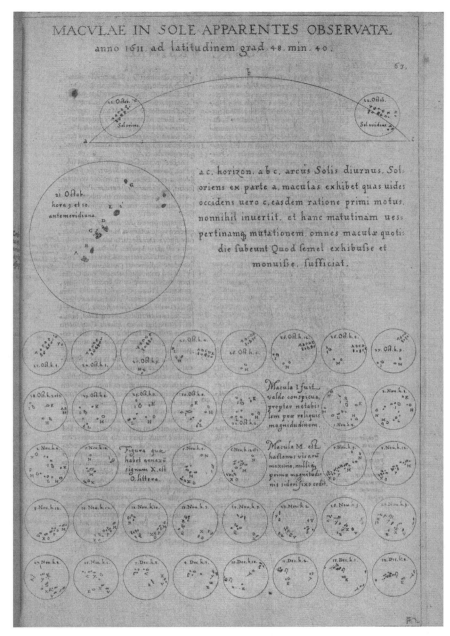

Figure 4.1. "Maculae in Sole Apparentes Observatæ", Christoph Scheiner, *Rosa Ursina sive Sol ex admirando facularum & macularum* (1630)

The method is this: Direct the telescope upon the sun as if you were going to observe that body. Having focused and steadied it, expose a flat white sheet of paper about a foot from the concave lens; upon this will fall a circular image of the sun's disk, with all the spots that are on it arranged and disposed with exactly the same symmetry as in the sun. The more the paper is moved away from the tube, the larger this image will become, and the better the spots will be depicted. Thus they will all be seen without damage to the eye, even the smallest of them—which, when observed through the telescope, can scarcely be perceived, and only with fatigue and injury to the eyes . . . Also, in order for the spots to be seen distinctly and with sharp boundaries, it is good to darken the room by shutting all the windows so that no light enters except through the tube . . . [*Letters*, 115–16]

In his recent important study, *Galileo's Instruments of Credit: Telescopes, images, secrecy*, Mario Biagioli identifies both the crucial innovation of this method—"as close as one could get to a mechanically produced image"—and its many advantages over Scheiner's method:

The advantages were many: no filters, a much higher level of detail, no need to limit observations to certain times of the day, no need to go back and forth between observing the Sun (in near-blinding conditions) and drawing from memory on a piece of paper (under very different lighting conditions), no problems with measuring the size and position of the spots or with maintaining the scale of the images constant, better visibility of weak sunspots, and minimization of the impact of personal drawing skills. It greatly routinized observation too.[39]

Biagioli characterized the effect of this method of observation and representation as the "de-skilling of sunspot observation" (p. 190) and notes the great advantage in the reproduction and dissemination of sunspot observations made possible by this entire methodology.[40]

One particularly interesting feature of this method works, in a sense, against the otherwise largely mechanical nature of projection: the starting point of the actual drawing of the projected image of the sun. Galileo begins the process by first drawing a perfect circle on the sheet of paper and then uses that circle—selected for wholly subjective if not arbitrary reasons—to orient the rest of the procedure. "I first describe on the paper a circle of the size that best suits me, and then by moving the paper towards or away from the tube I find the exact place where the image of the sun is enlarged to the measure of the circle I have drawn" (*Letters*, 115). Galileo states his perfectly rational rationale for this sequencing of the events that lead to the evidential figure he ultimately will draw: having first drawn the perfect circle on the paper it can then serve "as a norm and rule for getting the plane of the paper right, so that it will not be tilted to the luminous cone of sunlight that emerges from

---

[39] Biagioli (2006: 190).
[40] Biagioli's book (especially Ch. 3: "Between risk and credit: Picturing objects in the making") is a major contribution to work on Galileo's astronomy in general, and his study of sunspots, in particular. Biagioli also offers a comprehensive and detailed discussion of Scheiner's sunspot work, both in his pamphlets published by Wesler and in his later work *Rosa Ursina sive Sol ex admirando facularum & macularum* (1630).

the telescope" (*Letters*, 115–16), thereby avoiding all other possible shapes (the oval, for instance) other than the perfectly circular. The perfectly spherical and perfectly circular sun of Aristotle and the Peripatetics has been replaced by the perfectly drawn circle on Galileo's writing paper.

Galileo's subsequent discussion of the mechanics required to trace the initial drawing onto a second piece of paper (and which reads rather like a description of the practices of printing with moveable type) completes the circuit of writing that invents sunspots as *textually* real. Noting that the spots will be projected onto the writing surface both inverted and reversed "for the rays intersect one another inside the tube before coming through the concave lens" (*Letters*, 116), Galileo describes the process for righting the image-text:

> [S]ince we draw [the spots] on the side of the paper facing the sun, we have the picture opposite to our sight, so that the right-to-left reversal is already effected. Only the upper and lower parts remain inverted, so if we merely turn the paper upside down and bring the top ones to the bottom, we have only to look through the transparency of the paper against the light, at the spots will be seen precisely as if we were looking directly at the sun. And in that aspect they are to be traced upon another paper, in order to have them properly drawn. [*Letters*, 116]

As this passage makes clear, sunspots emerge materially into early modern scientific consciousness as "things" rendered material—"objects" rendered "real"—first through the use of the telescope as simultaneously an observational tool and writing instrument. Galileo's observational machine literally inscribes the fact of sunspots onto the writing surface. This action is not (merely) an act of reproduction, in large part because unlike, say, the modern photocopier and similar machines that mass produce copies of existing original texts, Galileo's machine produces *as image-text* an otherwise altogether non-reproducible *original*. His machine, in other words, while not fully independent or literally creative, is nevertheless instrumental to the textualization of sunspots, which otherwise remain (perhaps forever) unknowable, out of our grasp, and (for all practical purposes) non-existent. The second step in the production of sunspots relies upon the technologies of illustration and print. And it is through these steps that Galileo describes the work of science proper: no longer construed as immaterial—as illusions of either the telescopic or the optical viewing apparatuses—sunspots are established as materially real. For once Galileo is able to use the telescope to project sunspots onto a writing surface in a controlled and regimented fashion—a feat made possible by the strict and careful use of the telescope and the writing surface within the machinic scenario of observation and accurate recording—he is then able to submit the sunspots thus projected to meticulous metrology, thereby demonstrating the shape, number, location, and motion of these spots.

But that sunspots exist is not the *conclusion* that Galileo reaches in *Letters*. For, indeed he had asserted as much *before* his book on sunspots, in his *Discourse on Bodies in Water*. The work of *Letters on Sunspots*—and in this regard it is fundamentally different from *Sidereus Nuncius*—is not to announce a discovery, nor to reveal an object (in this case, a blemish) where before there seemed to be nothing, as was the case with the Medicean stars. But the sunspots emerge through the *Letters* not as new objects that emerge from the empty darkness of space (as did the Jovian moons), but instead as objects transformed from other objects. In Galileo's hands, sunspots are transformed from meandering swarms of tiny planets (as Scheiner had claimed) into objects that rest *demonstrably* on the surface of the sun.

Letters also differs fundamentally from the discourse of discovery on the level of method: where the starry message was essentially a matter of observation and narration, the sunspot book is essentially a matter of mathematics and representational proofs. Having produced sunspots on his drawing paper, Galileo is able to subject the sunspots to his mathematical machinations that ultimately serve to prove their nature—their location, relative size, motion, birth, and life and death on the surface of the sun:

> [O]ne must observe the apparent travel of the spots day by day. The spaces passed by the same spot in equal times becomes always less as the spot is situated nearer the edge of the sun. Careful observation shows also that these increases and decreases of travel are quite in proportion to the versed sines of equal arcs, as would happen only in circular motion contiguous to the sun itself. In circles even slightly distant from it, the spaces passed in equal times would appear to differ very little against the sun's surface. [*Letters*, 108][41]

And it is this set of characteristics, especially the matter of their movement across the disk of the sun, that in turn serves as the machine by which Galileo can determine—as well as demonstrate—the sphericity of the sun, together with its rotation and, ultimately, its central location within the orbits of its surrounding planets.[42]

But it would not be an overstatement to suggest that all of this theorizing depended upon the projection of the image of the sun by way of the use of the telescope as a writing instrument. Indeed, the observations Galileo mentions as decisive in determining motion on the solar disk were in fact observations of differences in his shadings on the writing paper from one moment of observation to another.[43] This is true even if it renders other statements in *Letters* problematical:

---

[41] Galileo's second letter also cites two other observational facts to support his conclusion: the appearance of the foreshortening of the sunspots as they depart from or draw closer to the rim of the sun, and the generally varying distance between spot along the same parallel as the approach to and move away from the center of the sun—"events" (as Galileo characterizes them) that "could be met with only in circular motion made by different points on a rotating globe" (Galileo [1957: 108]).

[42] As Drake argues as well, it was the mathematical proofs of the location and motion of sunspots that played a critical role in determining the nature and centrality of the sun—a feat that was accomplished (as Drake writes) when Galileo "began to apply terrestrial physics to the heavens" (*GAW*, 185). Drake then quotes, at length, the passage from *Letters* in which Galileo argues in favor of the idea that the sun itself actually rotates—and not merely what we might call an "atmosphere" that itself rotates around the otherwise stationary sun. This passage, as Drake writes in a footnote in *Discoveries*, is also notable for its discussion of the nature of motion: "The importance of this paragraph to the history of modern physics cannot be exaggerated." Drake continues:

> [W]hat it contains is the first announcement of the principle of inertia, according to which a body will preserve a state of uniform motion or of rest unless acted upon by some force. [Galileo (1957: 113)]

[43] Drake writes, "Without the aid of diagrams drawn by Castelli's method it had not been possible to measure sunspot motions accurately enough to be sure of this" (*GAW*, 185).

the assertion, for instance, that "all human reasoning must be placed second to direct experience" (*Letters*, 118). Or rather, it is in fact the "direct experience" of ink on white paper—together with their subsequent study and mathematical analysis—that is substituted for the direct experience of nature itself, which is rendered only *figurally* present in Galileo's science.[44]

A nature that is only figurally available to science is indeed an idea that is fundamental to Galileo and his astronomical work. And while there are many significant instances of this in his works, it is in the *Dialogue* that Galileo offers what is perhaps his most complete—and most powerful—formulation of this belief. On their third day of conversation, and in particular in response to Simplicio's description of a heliocentric universe (as theorized by Pythagoras), Sagredo asks, with some surprise:

> If this very ancient arrangement of the Pythagoreans is so well accommodated to the appearances, [one] may ask (and not unreasonably) why it has found so few followers in the course of the centuries; why it has been refuted by Aristotle himself, and why even Copernicus is not having any better luck with it in these latter days.[45]

In reply, Salviati offers a clarification—on the true nature of the surprise—and an extended explanation that points directly to the profound act of intellectual will required to place reason before sense (or direct) experience. "No, Sagredo," Salviati begins, "my surprise is very different from yours. You wonder that there are so few followers of the Pythagorean opinion, whereas I am astonished that there have been any up to this day who have embraced and followed it." He continues:

> Nor can I ever sufficiently admire the outstanding acumen of those who have taken hold of this opinion and accepted it as true; they have through sheer force of intellect done such violence to their own senses as to prefer what reason told them over that which sensible experience plainly showed them to the contrary. For the arguments against the whirling of the earth which we have already examined are very plausible, as we have seen; and the fact that the Ptolemaics and Aristotelians and all their disciples took them to be conclusive is indeed a strong argument of their effectiveness. But the experiences which overtly contradict the annual movement [of the earth in its orbit around the sun] are indeed so much greater in their apparent force that, I repeat, there is no limit to my astonishment when I reflect that Aristarchus and Copernicus were able to make reason so conquer sense that, in defiance of the latter, the former became mistress of their belief. [*Dialogue*, 327–8]

Taken together, these assertions on the nature of sense or direct experience and the intellectual work of reason that is necessary effectively (and even violently) to defeat the work of the senses help to secure the artifactual nature of Galilean

[44] In the *Dialogue*, as part of his ongoing critique of blind adherence to the works of Aristotle, Salviati's warning to Simplicio that philosophical arguments "must relate to the sensible world and not to one on paper" is similarly destabilized in the actual practices of Galileo's science (Galileo [1967: 113]).
[45] Galileo (1967: 327).

astronomical evidence and what will emerge in his writings as astronomical fact. There are, then, two acts of artifaction. The first is enabled by the telescope used as both an observational tool and as a writing instrument and that results in what we can confidently call Galileo's *production* of sunspots—a series of events made possible by Galileo's relocation of the universe within the telescope. In this, Galileo realizes what Cigoli reported Clavius to have said concerning telescopic observation of the heavens: that in order to see them—in order, in fact, for them to exist—it is first necessary to fashion a spyglass that can put them there. We can see a confirming instance of this when we return to Galileo's parable of the church window in which he claims that it is "nature" itself that demonstrates (and *has always demonstrated*) Galilean astronomical truths:

> I have since been much impressed by the courtesy of nature, which thousands of years ago arranged a means by which we might come to notice these spots, and through them to discover things of greater consequence. For without any instruments, from any little hole through which sunlight passes, there emerges an image of the sun with its spots, and at a distance this becomes stamped upon any surface opposite the hole. [*Letters*, 116–17]

Sunspots are made visible and manifest by virtue of nature itself; but we must remember Galileo's sleight-of-hand here, since this "nature" is itself already rendered artifactual by virtue of its constructedness. His narrative continues immediately:

> If in church some day Your Excellency sees the light of the sun falling upon the pavement at a distance from some broken windowpane, you may catch this light upon a flat white sheet of paper, and there you will perceive the spots. [*Letters*, 117]

The second act of artifaction concerns sense experience itself. Prior to Galileo's telescope-machine, sense experience is merely a matter of common sense, even if it is often wrong: the geostatic model of the universe accords perfectly with primary sense experience. Galileo's telescope-machine reinvents sense experience as the by-product of a set of technical operations. Once sense experience is transformed into an artifact of machines and their machinations, sense experience is no longer a matter of mere common sense (or popular opinions founded upon primary perception). Instead, a new kind of sense experience as an instrumentalized rationality triumphantly enters the world and it is this new form of sense experience—indeed, this new and transformed perceptual body—that emerges as both the culmination of Galilean astronomical science, and as its foundation.

For it is perfectly clear that Galileo's entire observational and textual practice of astronomical science both reinvents sense experience and then requires the verification of his discoveries and theories that only this new sense experience can provide. For Galileo, then, it can be said that belief comes before sight. Otherwise, the telescope merely frames empty and meaningless and dark space. The production of sunspots allows us to construct a counter-narrative to the conventional understanding of the relation between vision and belief that is said to typify or to define the so-called scientific method. Rather than serving as an instance of sight conferring belief ("seeing is believing"), the invention of sunspots serves to demonstrate the ways in which belief is first required in order that there could be a confirming vision.

5

John Donne's New Science Writing

One of the very earliest English responses to Galileo's *Sidereus Nuncius* appears in Donne's controversialist satire on the Jesuits, *Conclave Ignati*, published anonymously in January 1611, and in an English version—under the title *Ignatius His Conclave*—a few months later (also anonymously and likely translated from the Latin by Donne himself).[1] That such an early reaction to Galileo's epochal work should come from the pen of John Donne should not be surprising: Donne was, after all, the early seventeenth-century English poet perhaps most keenly interested in early modern science. At the same time, both as a poet and as a thinker, Donne was deeply conflicted about the science, especially the new astronomy, that had so caught his attention and that seems so profoundly to have complicated his understanding of the world. The nature of Donne's responses to the new science varied from work to work; in *Ignatius*, Donne's satire extends from his main target—the Jesuits and the culture of Catholicism more generally—to include the new science.

Ignatius begins with an act of disembodiment: the narrator falls into an "Extasie" in which his "little wandring sportful Soule" leaves his body with "liberty to wander through all places, and to survey and reckon all the roomes, and all the volumes of the heavens, and to comprehend the situation, the dimensions, the nature, the people, and the policy, both of the swimming Ilands, the *Planets*, and of all those which are fixed in the firmament" (*Ignatius*, 5–7). But rather than embark upon an account of these celestial wonders or begin a commentary upon them, the narrator instead offers a curious silence and a single marginal reference: "*Nuncius sydereus.*" What follows next is a pointed, if ironic, explanation of his refusal to take up the astronomical vision: "I thinke it an honester part as yet to be silent, then to do *Galilaeo* wrong by speaking of it, who of late hath summoned the other worlds, the Stars to come nearer to him, and give him an account of themselves" (*Igantius*, 7). The narrator similarly extends this right of priority to another great contemporary astronomer, Johannes Kepler "who (as himselfe testifies of himselfe) *ever since* Tycho Braches *death, hath received it into his care, that no new thing should be done in heaven without his knowledge. For by the law, Prevention must take place; and therefore what they have found and discovred first, I am content they speake and utter first*" (*Ignatius*, 7).[2]

[1] Donne (1969). For a discussion of the publication history of Donne's satire, see Healy's Introduction, xi–vii. Healy's edition contains both the Latin and the English versions of the text. All references in this chapter to *Ignatius His Conclave* are to this edition.

[2] Kepler himself responded (albeit indirectly) to his appearance in Donne's satire: "I suspect that the author of that impudent satire, the *Conclave of Ignatius*, had got hold of this little work [Kepler's

Having granted his astronomical contemporaries the privilege of their descriptions of the universe, however, the narrator does not pass by Galileo and Kepler without passing on to them at least one of his own observations: "Yet this they may vouchsafe to take from me," he declares, "that they shall hardly find *Enoch*, or *Elias* any where in their circuit" (*Ignatius*, 7). As T. S. Healy points out in his commentary on *Ignatius*, this "cryptic" remark refers to a debate over the identity of the two prophets of Revelations who will wage war against the Antichrist in the end times (*Ignatius*, 102). The specific identities of these unnamed prophet warriors were conventionally taken to be the Old Testament figures Enoch and Elias. In his own controversialist tract of 1608, *Triplici Nodo, Triplex Cuneus. Or an Apology for the Oath of Allegiance*, written (in part) in response to two papal breves issued by Paul V and Cardinal Bellarmine's famous letter to George Blackwell remonstrating the prominent secular priest for taking the Oath, King James argues against this convention, citing (among other pieces of evidence) the fact that Enoch and Elias are already in heaven and therefore unavailable for armed conflict on earth at the end of times.[3] The point, then, of the *Ignatius* narrator is that whatever may be the nature of Galilean or Keplerian astronomical observation and the universe thereby conceived, neither star gazer was likely to find the formidable Old Testament prophets floating between earth and heaven. Although clearly offered as ironic within the context of Donne's anti-Jesuit satire, and therefore intended in part as a piece of comedy, this brief moment certainly points to one of the many serious and contested points of interpretation—both theological and political—at stake in the controversialist exchanges between numerous English Protestant writers and their Catholic (and frequently Italian) counterparts in the early decades of the seventeenth century. Indeed, Donne's narrator continues this line of satiric thought. After having "surveid al the Heavens" and after seeing "all the rooms in Hell open to my sight," he remarks the virtually miraculous extension of his sight:

> And by the benefit of certaine spectacles, I know not of what making, but, I thinke, of the same, by which *Gregory* the great, and *Beda* did discerne so distinctly the soules of their friends, when they were discharged from their bodies, and sometimes the soules of such men as they knew not by sight, and of some that were never in the world, and yet they could distinguish them flying into Heaven, or conversing with living men, I saw all the channels in the bowels of the Earth; and all the inhabitants of all nations, and of all ages were suddenly made familiar to me. [*Ignatius*, 7–9]

The general target of this satiric passage is what Healy calls the "absurdities of the martyrologies" (*Ignatius*, 103) that were themselves the target of many Protestant controversialists, especially the idea that souls separated from bodies at the moment of death could appear to people left behind.[4] By implication, I would suggest, the satiric critique of the "spectacles" that allow the narrator's vision to penetrate into

Somnium], for he pricks me by name in the very beginning" (quoted in Nicolson [1956: 67]). Nicolson suggests that Kepler's dream book was in fact an influence on Donne's *Ignatius*.

[3] *Triplici Nodo* is reprinted in King James VI and I (1994: 85–131).

[4] See Healy for other examples of similarly skeptical moments in Donne's writing, including "The Resurrection," as well as the Sermons (Donne [1969: 103–4]).

the "channels in the bowels of the Earth" and lay all of Hell open to his sight, is also meant to discredit those "spectacles" through which the discoveries of *Siderus Nuncius* were made possible: the telescope. As Donne's narrator concludes his brief description of what his "spectacles" allow him to see ("all the rooms in Hell" and "all the inhabitants of all nations, and of all ages"), he points to other instances (likewise Catholic in nature) of perception aided by such mystical spectacles:

> I thinke truely, *Robert Aquinas* when he tooke *Christs* long Oration, as he hung upon the Crosse, did use some such instrument as this, but applied to the eare: And so I thinke did he, which dedicated to *Adrian* 6, that Sermon which *Christ* made in prayse of his father *Joseph*: for else how did they heare that, which none but they ever heard? [*Ignatius*, 9][5]

The purely fantastical nature of these instruments (unlike the telescope, which at least had the virtue of actual existence as an instrument) makes clear that the target of the satire and the critique in this passage is not new instruments as such, but rather any claim to extraordinary perception, especially if that perception is in some way sensory and if it has—or may have—any theological consequence. So Galileo's new universe, for instance, is enabled by a new kind of sensory perception, and though Galileo himself vigorously denied that there were any heretical implications that followed upon his discoveries, it was precisely the reputedly theological ramifications of them that led to such serious conflict with the Church. In *Ignatius*, Donne's narrator is (in a manner of speaking) used to what can be called a Catholic *methodology* that increases the likelihood that Galileo's claims will be similarly specious.

At the same time, the narrator informs us that Galilean astronomical discovery is but one instance of a species of sin: innovation itself. As it so often is in Donne's work, innovation here is an unequivocally negative feature of a great many human endeavors and the source of moral and religious error par excellence. After passing over the "*Suburbs* of Hel (I meane both *Limbo* and *Purgatory*) . . . so negligently, that I saw them not" (*Ignatius*, 9), the narrator proceeds to "more inward places" and sees "a secret place" to which "onely they had title, which had so attempted any innovation in this life, that they gave an affront to all antiquitie, and induced doubts, and anxieties, and scruples, and after, a libertie of beleeving what they would; at length established opinions, directly contrary to all established before" (*Ignatius*, 9). As his passing over the "Suburbs of Hel" makes clear, the architecture of the theological universe is the source of controversy. "*Purgatory* did not seeme worthy to me of much diligence," the narrator reports, "because it may seeme already to have been believed by some persons, in some corners of the *Romane Church*." This belief in the existence of Purgatory emerges here as the consequence of the fact that the Catholic Church itself is manifestly guilty of innovation.

---

[5] Donne prints two titles in the margin at this point: Alphonso Paleoti's *Stigmata Sacrae Sindoni Impressa* and Gieronimo Gracian's *Josephina: Summario de las Excelencias del Glorioso S. Joseph*; see Healy's explanatory note (*Ignatius*, 104–5).

Catholics have believed in Purgatory "for about 50 yeares," the narrator writes, "ever since the Councell of *Trent* had a minde to fulfill the prophecies of *Homer*, *Virgil*, and the other *Patriarkes* of the *Papists*; and being not satisfied with making one *Transubstantiation*, purposed to bring in another: which is, to change *fables* into *Articles* of faith" (*Ignatius*, 9).

The attempt here to redirect the very charge of dangerous and heretical innovation levied against the Protestant Reformation onto the Catholic Church is standard fare in controversialist discourse of the period and *Ignatius* is no different in this regard.[6] But *Ignatius* is characterized by an undisguised joy and playfulness in the articulation of its critique of Catholic and especially Jesuit innovation. Having progressed to the "secret place" within the heart of hell, where he sees both Pope Boniface III and "*Mahomet*" who "seemed to contend about the highest roome. Hee gloried of having expelled an old Religion, and *Mahomet* of having brought in a new" (*Ignatius*, 9), the narrator will stand witness to a virtual parade of "pretenders" (*Ignatius*, 13) for the place of honor nearest to Lucifer.[7] The criteria for consideration are clear:

> Now to this place, not onely such endeavour to come, as have innovated in matters, directly concerning the soule, but they also which have done so, either in the Arts, or in conversation, or in any thing which exerciseth the faculties of the soule, and may so provoke to quarrelsome and brawling controversies: For so the truth be lost, it is no matter how. (*Ignatius*, 13)

Significantly, the first pretender is "a certaine *Mathematitian*" who "with an erect countenance, and setled pace, came to the gates, and with hands and feet (scarce respecting *Lucifer* himelfe) beat the dores, and cried, 'Are these shut against me, to whom all the Heavens were ever open, who was a Soule to the Earth, and gave it motion?'" (*Ignatius*, 13). The narrator quickly identifies this figure as Copernicus, whose boasting grows even more extravagant when Lucifer asks him about the nature of his claims for innovation. Some of Copernicus's innovations were, he claims, explicitly to the benefit of Lucifer himself: "I am he, which pitying thee who wert thrust into the Center of the world, raysed both thee, and thy prison, the Earth, up into the Heavens."[8] But Copernicus saves his greatest claim for his last:

---

[6] For an important discussion of Donne's contributions to controversialist literature, including, in particular, his 1610 *Pseudo-Martyr*, see Sommerville (2003). See also, Cain (2006).

[7] Donne does not leave this contest undecided.

> But it is to be feared, that *Mahomet* will faile therein, both because hee attributed something to the old *Testament*, and because he used *Sergius* as his fellow-bishop, in making the *Alcoran*; whereas it was evident to the supreme Judge *Lucifer*, (for how could he be ignorant of that, which himselfe had put into the Popes mind?) that *Boniface* had not onely neglected, but destroyed the policy of the State of *Israel*, established in the old *Testament*, when he prepared *Popes* a way, to tread upon the neckes of *Princes* . . . [*Ignatius*, 9–11]

[8] "By my meanes," Copernicus claims, "*God* doth not enjoy his revenge upon thee." This follows, he argues, as a direct consequence of his reorganization of the universe: "The Sunne, which was an officious spy, and a betrayer of faults, and so thine enemy, I have appointed to go into the lowers part of the world" (*Ignatius*, 15).

Shall these gates be open to such as have innovated in small matters? and shall they be shut against me, who have turned the whole frame of the world, and am thereby almost a new Creator? [*Ignatius*, 15]

Lucifer is "stuck in meditation" for a response, while for his part Ignatius Loyola ("a subtile fellow") is "so indued with the Divell, that he was able to tempt, and not onely that, but (as they say) even to possesse the Divell, apprehended this perplexity in *Lucifer*" (*Ignatius*, 15) and intervenes decisively against Copernicus and his claims:

But for you, what new thing have you invented, by which our *Lucifer* gets any thing? What cares hee whether the earth travell, or stand still? Hath your raising up of the earth into heaven, brought men to that confidence, that they build new towers or threaten God againe? Or do they out of this motion of the earth conclude, that there is no hell, or deny the punishment of sin? Do not men believe? do they not live just, as they did before? [*Ignatius*, 17]

In fact, Ignatius suggests that if Copernicus's theory is right ("that those opinions of yours may very well be true"), then he literally has no claim, either for innovation, or for the place of honor alongside Lucifer. "If therefore any man have honour or title to this place in this matter, it belongs wholly to our *Clavius*, who opposed himselfe opportunely against you, and the truth, which at that time was creeping into every mans minde" (*Ignatius*, 17).[9] In this argument that both ends Coperni-cus's claims in hell and stands as an astonishingly prescient reading of the very doctrinal debates that will take center stage in Galileo's conflicts with the Catholic Church in the early 1630s, Donne's Ignatius refutes Copernicus by noting that the Church had never codified *as doctrine* the Ptolemaic universe that early modern astronomy will demolish—the only act that could render Copernican–Galilean discovery heretical:

Let therefore this little *Mathematitian* (dread Emperour) withdraw himelfe to his owne company. And if heereafter the fathers of our Order can draw a *Cathedrall Decree* from the Pope, by which it may be defined as a matter of faith: *That the earth doth not move*, & an *Anathema* inflicted upon all which hold the contrary: then perchance both the Pope which shall decree that, and *Copernicus* his followers, (if they be Papists) may have the dignity of this place. [*Ignatius*, 19][10]

Lucifer agrees and Copernicus, "without muttering a word, was as quiet, as he thinks the sunne" (*Ignatius*, 19), is forever silenced. Thus ends the attempt of the

---

[9] Ignatius continues his repudiation of Copernicus as innovator by pointing to classical precedents for a heliocentric universe:

But your inventions can scarce bee called yours, since long before you, *Heraclides, Ecphantus*, & *Aristarchus* thrust them into the world: who notwithstanding content themselves with lower roomes amongst the other Philosophers, & aspire not to this place, reserved onely for *Anti-Christian Heroes*: neither do you agree so wel amongst your selves, as that you can be said to have made a *Sect*, since, as you have perverted and changed the order and *Scheme* of others: so Tycho Brachy hath done by yours, and others by his. [*Ignatius*, 19]

[10] Of course, this does happen with the Office of the Inquisition's decree against Galileo.

first pretender, and thereby is set the model for all subsequent pretenders—Paracelsus, Machiavelli, Aretino, Columbus, and Philip Neri—who are all defeated by the greater claims to dangerous and heretical innovation by Ignatius Loyola himself.

Donne brings Galileo and his astronomical machines back in the culminating scene of the book in which he plays out a comic fantasy of punishment for innovators and for Ignatius Loyola in the form of a great excommunication. Lucifer declares Ignatius the winner and announces his plan to enshrine him in a new hell:

> But since I may neither forsake this kingdome, nor divide it, this onely remedy is left: I will write to the Bishop of *Rome*: he shall call *Galilaeo* the *Florentine* to him; who by this time hath throughly instructed himselfe of all the hills, woods, and Cities in the new world, the *Moone*. And since he effected so much with his first *Glasses*, that he saw the *Moone*, in so neere a distance, that hee gave himself satisfaction of all, and the least parts in her, when now being growne to more perfection in his Art, he shall have made new *Glasses*, and they received a hallowing from the *Pope*, he may draw the *Moone*, like a boate floating upon the water, as neere the earth as he will. And thither . . . shall all the Jesuites bee transferred, and easily unite and reconcile the *Lunatique Church* to the *Romane Church*; without doubt, after the Jesuites have been there a little while, there will soone grow naturally a *Hell* in that world also: over which, you *Ignatius* shall have dominion, and establish your kingdome & dwelling there. [*Ignatius*, 81][11]

The skepticism on display here for both the objects of astronomical discovery and the instrument (or *machine*) that enables them is a familiar feature of reactions to Galileo's celestial observational works—whether such reactions are relatively early (such as Donne's) or rather later (as one sees, for instance, in the scientific and creative works of Margaret Cavendish in the 1660s and beyond).[12] Donne's skepticism in *Ignatius* takes the form of satire and the tone of mild comedy, even though both are informed, I would argue, by a more serious understanding of the serious nature of the new science—and the serious nature of both the theological and the social implications that the new science was thought to precipitate. As such, Donne's inclusion of such matters as the nature of embodiment, for example, or the contested relationship between community (whether religious or social) and isolation (cast in the model of excommunication), are especially noteworthy and, in fact,

---

[11] Lucifer continues to outline the eventual ascendency of Ignatius: "And with the same ease as you pass from the earth to the *Moone*, you may pass from the *Moone* to the other *starrs*, which are also thought to be worlds, & so you may beget and propagate many *Hells*, & enlarge your *Empire*, & come nearer unto that high seate, which I left at first" (*Ignatius*, 81). As this vision suggests, Donne may have in mind as target here not Galileo (who never argued that other stars were worlds), but instead Giordano Bruno, who certainly did.

[12] Cavendish's rejection of Hooke's work, along with the operation and use of the microscope, constitutes a striking instance of her rejection of the instrumental observational dimension of experimental science in *Observations*; see I.3 "Of Micrography, and of Magnifying and Multiplying Glasses" in Cavendish (2001), which includes the following summary conclusion:

> Wherefore the best optic is a perfect natural eye, and a regular sensitive perception; and the best judge, is reason; and the best study, is rational contemplation joined with the observations of regular sense, but not deluding arts; for art is not only gross in comparison to nature, but, for the most part, deformed and defective, and at best produces mixt or hermaphroditical figures, that is, a third figure between nature and art . . . [p. 53]

constitute a deep engagement with the new science that seems, at least on the level of theme, very easily ridiculed and satirized in *Ignatius*.

The nature of embodiment (what is the soul, how precisely does it relate to the body, what is its fate upon the death of the body?), together with concerns about what may emerge as a surprisingly fragile community in the face of the threat of its fragmentation (figured comically in Igantius Loyola's excommunication to another world), are manifestly weighty and timely matters for Donne, and for the seventeenth century more generally, and frequently engendered serious and demanding responses from the poet. Within a few months of the composition and publication of *Ignatius*, Donne returned to this constellation of concerns in "An Anatomy of the World: The First Anniversary," in which he offers this memorable, somber, and perhaps even grave formulation:

> And new philosophy calls all in doubt,
> The element of fire is quite put out;
> The sun is lost, and th'earth, and no man's wit
> Can well direct him, where to find it. ["First Anniversary," ll. 205–8][13]

Typical of Donne's responses to the new science, this passage points to the disruptive nature of the new science and its many destabilizing discoveries. At the same time, Donne does not explicitly reject or accept either the science or its new knowledge; rather, the poet nervously inhabits the newly destabilized world, if only long enough to meditate on a number of the immediate consequences that might follow from the rupture of the established and traditional world effected by new science. "The First Anniversary" continues in just this mood:

> And freely men confess, that this world's spent,
> When in the planets, and the firmament
> They seek so many new; they see that this
> Is crumbled out again to his atomies.
> 'Tis all in pieces, all coherence gone;
> All just supply, and all relation:
> Prince, subject, father, son, are things forgot,
> For every man alone thinks he hath got
> To be a phoenix, and that there can be
> None of that kind, of which he is, but he. [ll. 209–18]

The new science announces its discoveries, but rather than marking something like the dawning of greater natural knowledge (of the celestial movements, for instance), these announcements signal instead the world winding down toward incoherence and dissolution. This is Donne's articulation of the entropic principle, *avant la lettre*: new scientific knowledge is the harbinger of increasing chaos. This is the fall into ever-increasing disorder: the emergence of new (celestial) objects serves only to fragment and to disintegrate the known world, even to the "atomization" of that previously whole and coherent and meaningful universe. Now memory fails,

---

[13] Donne, "An Anatomy of the World: The First Anniversary," in Donne (1990: 212).

coherence is utterly lost and the social result is a kind of universal excommunication: the destruction of community through the radical isolation of each "man," phoenix-like, into a pure and abject individual. This version of the "new philosophy" and the doubt it introduces into the suddenly fractured, fragmented, and "atomic" world is entirely consonant with the theme of the sick, lame, halting, and, finally, dead world anatomized in "The First Anniversary."

The poem was written by Donne in 1611 to honor the memory of Elizabeth Drury, the recently deceased daughter of Sir Robert Drury, one of Donne's principal patrons, and is one of several important funeral poems (which were something of a specialty) that Donne wrote across his career.[14] The presiding conceit of the poem is that the untimely death of Elizabeth Drury is a manifestation of a generalized decay that afflicts the entire world and all the people in it: "She, she is dead; she's dead: when thou knows't this,/Thou knows't how lame a cripple this world is" (ll. 237–38). And what further knowledge do we gain from the poem that has undertaken the analytical task of the anatomy of the world? Little knowledge, the poem says, and even less comfort:

> That this world's general sickness doth not lie
> In any humour, or one certain part;
> But, as thou saw'st it rotten at the heart,
> Thou seest a hectic fever hath got hold
> Of the whole substance, not to be controlled,
> And that thou hast but one way, not to admit
> The world's infection, to be none of it.
> For the world's subtlest immaterial parts
> Feel this consuming wound, and age's darts. [ll. 240–8]

On the level of theme, then, we might say that "The First Anniversary" in part conforms to a traditional funereal or elegiac *topos*, with, significantly, the added feature of the indictment of the new science as yet one further symptom of universal collapse and decay. But on another level, "The First Anniversary" enacts a quite unusual and unconventional series of maneuvers that both push against the funereal/elegiac *topos* and, at the same time—and perhaps surprisingly—move toward an embrace of a particular aspect or dimension of the new science that I want to call the new science writing.

"The First Anniversary" identifies the death of Elizabeth Drury as an instance—as an *example*—of larger truths about the material world that the poem is careful to inventory in elaborate (and perhaps obsessive) detail. But, in spite of the poem's litany of the fallen and corrupt nature of the world, it is also clear that the Elizabeth Drury of the poem is a person only and entirely in the abstract. As is well known, Donne had never actually met her and his praise of her virtues is therefore either simply conjectural, or it is formulaic, the result, as it were, of the generic demands

---

[14] For a discussion of Donne's relationship with Sir Robert Drury (who as patron "directly or indirectly exercised a deeper influence on his life and work than any of the others," see Bald (1970: 237–62).

of the funeral poem as such.[15] The poem, then, rather than working as a first-hand testimonial of the truths it speaks about Elizabeth Drury as a person, is in fact resolutely impersonal and the identity it ostensibly means to commemorate (in loss) is a strangely absent or disembodied one. As such, the poem's only true subject is in reality the narrator; Elizabeth Drury emerges if at all (her name is never spoken, we notice, and the poem provides no specificity in its account of its ostensible subject) only as a screen onto which the narrator projects his own (poetic) identity. This is the difference *on the level of method* (rather than on the level of theme or subject matter) that is linked to the logic of the early modern science and that I will examine in this chapter: the movement toward the discourse of exemplarity. Keeping in mind that the only subject that can be said to be embodied in the poem is the narrator, it follows that the experience reflected in the poem can hardly be Elizabeth Drury's. Rather, it is the poet's own. The poem thus puts into play an innovation in Donne's work on the level of method that pivots on the self-reflexive and autobiographical account of embodied experience that secures both poetic and social meaning to the extent that it is made to emerge as exemplary.

Among the many interpretive things that can be said about "The First Anniversary"—that it is a religious poem, that it is a funeral poem, that it represents Donne's experimentation with an essentially Catholic meditative form—we can also say that it is a lamentation over a kind of excommunication figured in the collapse of community into isolated monads, the atomic man cast in light of the new philosophy's entropic push toward fragmentation. But at the same time, and in a contrary movement, the poem also embodies the narrator's realization that the poem, like the death of Elizabeth Drury, in fact, functions best when understood in relation to the speaking self. And that this speaking self, moreover, functions best once it is made to fulfill the demands of the discourse of exemplarity. As I will argue in the following pages, the mechanism that allows exemplarity to succeed is Donne's creative and skeptical engagement with the structure of early modern science which is itself similarly dedicated to the logic of the exemplary.

It is this process that I would like to trace in this chapter. In order to do so, in order to rethink the issues that are fundamental to Donne's poetic imagination—in particular, the nature of embodiment and the importance of the autobiographical to Donne's project of exemplarity—it will be necessary to reconsider Donne's

---

[15] In a letter "To Sir G.F." in Donne (1651, rpt. 1977), Donne responds to comments about the poem he had reached him during his trip to the continent with Sir Robert Drury:

> I hear from *England* of many censures of my book, of Mistress *Drury*; if any of those censures do but pardon me my descent in Printing any thing in verse, (which if they do, they are more charitable then my self; for I do not pardon my self, but confesse that I did it against my conscience, that is, against my own opinion, that I should not have done so) I doubt not but they will soon give over that other part of that indictment, which is that I have said so much; for no body can imagine, that I who never saw her, could have any other purpose in that, then that when I had received so very good testimony of her worthinesse, and was gone down to print verses, it became me to say, not what I was sure was just truth, but the best I could conceive; for that had been a new weaknesse in me, to have praised any body in printed verses, that had not been capable of the best praise that I could give. [pp.74–5]

Gosse (1959) dates this letter to April 1612 (p. 307).

relationship to the new science. As I outlined briefly in the first chapter of this book, Donne's engagement with early modern science has long been an interest of literary critics (though not always in the same ways or for the same purposes). In the following section of the chapter, I pursue the investigation of Donne's relation to emergent science by returning to the work on the poet by one such critic—William Empson—whose understanding of Donne is especially provocative within these contexts, particularly his argument concerning Donne's *non-metaphorical* relation to the new philosophy, and in part for his argument about the place of the (auto) biographical in literary critical method.

In the next section of this chapter I turn to a sustained reading of another of Donne's texts that is dedicated to the sustained examination (this time in prose) of some of the issues that are central to "The First Anniversary," including embodiment, isolation, exemplarity, and the ravages of disease on the body and the community: Donne's 1623 *Devotions upon Emergent Occasions*. I will be particularly interested to examine the method by which Donne's embodied experience is made to signify as generalized and exemplary within the practice of the new science writing. This section begins with a discussion of the *Devotions* in relation to early modern autobiography and the question of materialism that is fundamental to both. In order to respond to these concerns, this section turns to an analysis of a group of three Meditations in the *Devotions* that collectively secure Donne's successful negotiation of the "problem" of embodiment and exemplary experience through a theorization of the tolling of the bells.

I conclude this chapter with a brief consideration of Donne's *Devotions* from the perspective of what in many ways constitutes its opposite—and antagonistic—resolution to the questions of embodiment, experience and knowledge. As I will argue, René Descartes offers just this kind of challenge to the *Devotions* in his establishment of the (philosophical) dualism that Donne's works so assiduously to pre-empt.

I

Donne's attention in *Ignatius* to Galileo—and to instrumental astronomical scientific observation more generally—has been duly noted by readers, some of whom have been interested in gauging his relation to the new science. We saw an important example of this in the work of Marjorie Nicolson, discussed in chapter one. While it is not my concern here to catalogue critical responses to Donne's perceived relationship to "new philosophy" in general, it is important to note the enduring nature of this concern—from Charles Coffin's *John Donne and the New Philosophy* (1937) through the work of Louis Martz and Frank Kermode in the 1950s, to R. C. Bald, John Carey, and George Parfitt in the 1970s and 1980s, to very recent studies, such as Angus Fletcher's *Time, Space, and Motion in the Age of Shakespeare* and Richard Sugg's *John Donne*, both in 2007.[16] In keeping with the

---

[16] For another discussion of Donne and early modern science, see Ch. 5 "New philosophy" in Sugg (2007: 123–53); for a discussion of Donne's *poetic* response to the potentially destabilizing

interests expressed in that opening chapter (and, indeed, throughout this book), my concern here is not to sketch the influence of early modern science on Donne's poetry and prose, but rather to examine the ways in which Donne's work shares certain features—epistemological and discursive in nature—with science. As such, my first question asks if Donne's interest in the new science is a metaphorical one. Or, is his attraction to the new science, as well as his interest in it, more a matter of *method*?

Among those critics interested in the issue of Donne and the new science, William Empson provides an interesting help for thinking about this question. And Empson does so in his typically arresting ways. In "Donne the Space Man," published in 1957 (the year of Sputnik), Empson offers this curious—because cosmic, because outlandish, and because rigorously playful—comment: "Donne, then, from a fairly early age, was interested in getting to another planet much as the kids are nowadays."[17] Empson understands, of course, the controversial nature of such an assertion. Or rather, he is aware that this assertion *has become* controversial: "Present-day writers on Donne," he notes, "have never heard of a belief about him which, twenty or thirty years ago, I thought was being taken for granted. I can't believe I invented it; it was part of the atmosphere in which I grew up as an undergraduate at Cambridge" (p. 78). This belief has never been refuted, Empson says, but neither has the evidence to support it been assembled—in large part because it has seemed unnecessary: "I myself, being concerned with verbal analysis, thought I could take this part of Donne's mind as already known; and all the more, of course, when I was imitating it in my own poems, which I did with earnest conviction" (p. 78). This, however, is no longer the case: "The current of fashion or endeavour has now changed its direction and a patient effort to put the case for the older view seems timely" (p. 78). With this, Empson sets out "to state the position as a whole" (p. 78), even in the face of certain objections and reservations:

> In our present trend of opinion, as I understand, to impute this belief to the young Donne will be felt to show a lack of sense of history, to involve a kind of self-indulgence, and to be a personal insult to the greater preacher. I shall be trying to meet these objections, but had better say at once that I think the belief makes one much more convinced of the sincerity of his eventual conversion, and does much to clear the various accusations which have recently been made against his character. [p. 79]

Empson understands that he runs the risk of seeming to claim something ludicrous for Donne—and for anyone at all, for that matter, at least as far as space travel is concerned. "No reasonable man," Empson assures us, "would want space travel as such; because he wants to know, in any proposal for travel, whether he would go farther and fare worse" (p. 79). Empson then offers an interesting biographical

---

implications posed by new science, see Fletcher (2007, especially Ch. 7 "Donne's apocryphal wit," pp. 113–29). Discussing the character of Donne's poetry, Fletcher speaks directly to its famous difficulty: "A poetry flexible enough to express and therefore handle cultural incoherence will need simultaneously to impose and to retract its own self-censorship, thereby producing a new mode of almost scientific, at times metaphysically tangled, apocryphal wit" (p. 128).

[17] Empson (1993: 78). Subsequent citations refer to this collection of Empson's work.

moment (interesting in part because it transforms into something of an autobio-graphical moment), as he pauses to consider space travel and one particular representative of "kids . . . nowadays":

> A son of my own at about the age of twelve, keen on space travel like the rest of them, saw the goat having kids and was enough impressed to say "It's better than space travel." It is indeed absolutely or metaphorically better, because it is coming out of the nowhere into here; and I was so pleased to see the human mind beginning its work that I felt as much impressed as he had done at seeing the birth of the kids. One does not particularly want, then, to have Donne keen on space travel unless he had a serious reason for it. [p. 79]

It is Empson's sense, of course, that Donne indeed had serious reason for the contemplation of the separate planet and in a number of essays—especially "Donne the Space Man"—articulates a long, complex, and, I believe, compelling discussion of the centrality of the discourse of the separate planet and the theology of the separate planet to Donne's literary works.[18] Citing Donne's status as "an adviser to Anglican officials" on matters of theological controversy, and his ambivalent relationship to his Catholic ancestry (and his family's illustrious Catholic martyrs), together with his having read Kepler's 1609 *De Stella Nova* and Galileo's 1610 *Sidereus Nuncius*, Empson writes:

> It is from this background that he was keenly, if sardonically, interested in the theology of the separate planet—from fairly early, though he did not come to feel he was actually planted on one till he realized the full effects of his runaway marriage. By the time he took Anglican Orders I imagine he was thankful to get back from the interplanetary spaces, which are inherently lonely and ill-provided. I don't deny that he was very capable of casuistry—his sense of honour would work in unexpected ways; but at least he had joined the only church which could admit the existence of his interplanetary spaces. [p. 84]

For Empson, Donne's interest in the separate planet—which changes from a more or less obvious and explicit fantasy in Donne's early work to an idea that Donne the cleric eventually comes to understand as something of a heresy, which he will cite (in the *Sermons*, for example) for the specific purpose of refuting—is one of the keys to understanding Donne's literary work. In one striking passage, for instance, in defense of a particular line of Donne's poetry against the charge of being "ugly," Empson writes: "The sound requires it to be said slowly, with religious awe, as each party sinks into the eye of the other; it is a space-landing" (pp. 96–7). The "space-landing" imagined here refers to the line from "The Good Morrow" that concludes the poem's second stanza:

> And now good morrow to our waking souls,
> Which watch not one another out of fear;

---

[18] Heffenden's collection of Empson's essays related to Donne—and questions of early modern cosmology more generally—includes a number of essays taken from typescript left unpublished at the time of Empson's death; these include "Copernicanism and the censor," "Thomas Digges his infinite universe," and "Godwin's voyage to the moon."

> For love, all love of other sights controls,
> And makes one little room, an everywhere.
> Let sea-discoverers to new worlds have gone,
> Let maps to others, worlds on worlds have shown,
> Let us possesse one world, each hath one, and is one.[19]

Empson writes:

> Then there is a pause for realisation, and the next verse begins in a hushed voice but
> with a curiously practical tone: "You know, there's a lot of evidence; we really are on a
> separate planet." It is never much use talking about a sound-effect unless you know
> what it is meant to illustrate. [p. 97]

On Empson's reading, the figure of the separate planet that appears across the range
of Donne's work, both before and after his conversion to Protestantism, is of central
importance and marks something more than a merely metaphorical significance in
Donne's literary imagination. In an earlier essay, "Donne and the rhetorical
tradition" (1949), Empson in fact had taken serious issue with the tendency he
found on full display in Rosemund Tuve's book *Elizabethan and Metaphysical
Imagery* (among others) for "the exercise of analyzing Donne's rhetoric ... to
'explain things away'"—a feature of Tuve's work that, as Empson saw it, served
only to reduce Donne's interplanetary interest to mere metaphor. "As I understand
her," Empson writes, "she treats the Donne line of talk that the idealised woman is
a world, or that the two happy lovers are a world, as a straightforward use of the
trope *amplificatio*." And this is precisely Empson's complaint:

> I do not think you get anywhere with Donne unless you realise that he felt something
> different about his repeated metaphor of the separate world; it only stood for a
> subtle kind of truth, a metaphysical one if you like, and in a way it pretended to be
> only a trope; but it stood for something so real that he could brood over it again and
> again.[20]

As suggested above, Empson's faith in the discourse of the separate planet is
intimately connected to the matter of Donne's "runaway" marriage—though,
certainly, this is but one of the ways in which the separate planet signifies in
Donne's literary work.[21] For Empson, the separate planet is fundamentally a
theological notion: "The young Donne, to judge from his poems, believed that
every planet could have its Incarnation, and believed this with delight, because it

---

[19] Donne, "The Good Morrow" in Donne (1990: 90, ll. 8–14).
[20] Empson (1993: 69).
[21] The matter of Donne's marriage is centrally important and leads to another memorable assertion
from the essay: "The hostess of the palace of the Keeper of the Great Seal did behave splendidly under
the terrible privations of life on a separate planet" (Empson [1993: 119]). The "hostess" in question is,
of course, Anne Donne and the "life on a separate planet" is meant to refer to the years of social
isolation—and economic deprivation—that followed upon her secret marriage to Donne, complete
with what was effectively an exile to Mitcham, and which was to become such an important and
standard feature not only in the biography of the poet, but also in subsequent interpretations of his
work, as well.

automatically liberated an independent conscience from any earthly religious authority" (p. 81).[22]

A significant portion of "Donne the Space Man" is given over the often elaborate discussion of occurrences of planets—or spheres—in a range of Donne's poems, with each, according to Empson, illustrating the underlying theological argument. Not all readers, however, have always found the image of the separate planet in some of these same poems. One importance instance in which Empson argues for the separate planet against a prevailing contrary argument concerns Donne's poem, "A Valediction: forbidding Mourning." Empson quotes the third stanza:

> Moving of th'earth brings harms and fears,
> Men reckon what it did and meant,
> But trepidation of the spheres,
> Though greater far, is innocent.[23]

As Empson points out, as early as Coffin's *John Donne and the New Philosophy* (1937), the "Moving of th'earth" was identified as a figure for the earth as planet; Empson quotes Coffin: "Of the new astronomy, the 'moving of the earth' is the most radical principle; of the old, the 'trepidation of the spheres' is the motion of the greatest complexity . . . Donne does not deny the truth of either doctrine—if he did the whole point of the argument would be wasted and futile" (Empson, 122).[24] But the critical case, Empson laments, has altered and this particular image in this poem has "become rather a victory" for what Empson calls "the new rigour":

> As time went on, this case became rather a victory for the new rigour, in its campaign to make poetry as dull as possible; Mr W. K. Wimsatt in his essay "The Intentional Fallacy" (first published in 1946) proved that the line refers to an earthquake. This would do actual *harm*, at a point of time, and afterwards (hence the past tense) men would reckon the damage and invent a superstitious *meaning*, none of which applies literally to the new astronomy. [p. 122][25]

Empson's distaste for Wimsatt (and the Wimsatt and Beardsley essay) is palpable—and especially meaningful in his own essay's discussion and examination of method and the place that ought to be accorded to the "private" in literary critical analyses. Discussing Wimsatt's 1954 book, *The Verbal Icon*, Empson writes, "Mr Wimsatt gave a qualifying sentence before the bang of his final one, but I expect he felt that only made the bang stronger" (p. 122), then quotes Wimsatt:

---

[22] And, in truth, Empson's separate-planet discussion is partly underwritten by his sense that critics writing in 1957 had no greater knowledge about other planets—and the question of other living inhabitants of them—than poets writing in the early modern period had: "When we study a man in the past grappling with a problem to which we have learned the answer we find it hard to put ourselves in his position; but surely a modern Christian knows no more about this than Donne or Bruno, and has no occasion for contempt" (Empson [1993: 81]).

[23] Donne, "A Valediction: forbidding Mourning" in Donne (1990: 120, ll. 9–12).

[24] Coffin (1937: 98).

[25] W. K. Wimsatt wrote "The intentional fallacy" with Monroe C. Beardsley and it was first published in the *Sewanee Review* in 1946; Empson quotes from Wimsatt (1954).

Perhaps a knowledge of Donne's interest in the new science may add another shade of meaning, an overtone to the stanza in question, though to say even this runs against the words. To make the geocentric and heliocentric antithesis the core of the metaphor is to disregard the English language, to prefer private evidence to public, external to internal.[26]

Empson laconically intones: "The way that *private* goes with *external* is rather subtle" (p. 123). Even if "earthquake" is the "chief meaning" (which Empson grants), this does not mean we must "deny that Donne...was conscious of a secondary meaning which he rather hoped to insinuate":

> The effect of giving the phrase both meanings is to say, "And also the sudden introduction of the idea that there may be life on other planets has affected the Churches like an earthquake." At this remove, the phrases can apply tidily to Copernicanism; the threat to the churches' absolutism, Donne can mean, has frightened them to persecution. [p. 123]

As Empson then makes explicit, he understands this as an instance of what he called an "equation" in *The Structure of Complex Words*, according to which definition—in the local example—"in the mind of Donne, the 'major sense' of 'moving of the earth' is the Copernican one, but the sense demanded by the immediate context is the earthquake."[27] He continues:

> indeed, this [i.e. the earthquake] is also what the reader is likely to take as the major sense, and yet Donne thinks of Copernicanism as the subject of his equation, the part that comes first in the sentence expounding it. That is, the Copernicanism, and not the earthquake, is the one you are expected to understand better after the two have been compared. [p. 123]

Empson points to other moments in the poem that help prepare the Copernican meaning of the figure in the poem. Pointing to the poem's second stanza, for instance:

> So let us melt, and make no noise,
> No tear-floods, nor sigh-tempests move,
> 'Twere profanation of our joys
> To tell the laity of our love. [ll. 5–8]

Empson argues that the "second half [of the stanza] takes for granted that they are a separate religion" while the first "though less obviously" locates the lovers on a separate planet: "It would be tiresome to use these conventional hyperboles, while the tone treats them as commonplace, without a moderating idea that the floods and tempests are only rightly so called within the private world of the lovers." This is another instance of a "situation [that] regularly suggested to him his secret planet, so naturally a Copernican idea came into his mind for the next verse." In this particular instance, however, Donne "did not much want it for this poem, so he thrust it down to a secondary meaning" (pp. 123–4). Empson then delivers the

---

[26] Wimsatt (1954: 14).   [27] Empson (1951).

crucial argument—but an argument that does not take the poem (or its figurations) as its focus, but rather addresses itself to the matter of critical method:

> If you dislike my claiming to know so much, I have to answer that I think it absurd, and very harmful, to have a critical theory, like Mr Wimsatt's, that a reader must not try to follow an author's mind. [p. 124]

The urgency in Empson's article derives not only from the strength of his belief in Donne's poetics and theology of the separate planet, but also—and for my purposes here, more importantly—from his desire to respond to criticisms of Donne *on the level of method*. As his critiques of the merely tropological reading method practiced by Tuve (for example) and the even more interesting rejection of the merely aesthetic method practiced by Wimsatt—and enshrined, in many ways, as indeed the core of New Criticism—make clear, Empson's method requires a more exhaustive (even if often speculative) engagement with the "author's mind."[28] In all practical terms, this means the biographical becomes central to Empson's critical method.

A second feature of Empson's method in "Donne the Space Man" is his insistence that one avoid reading Donne's engagement with the discourse of the separate planet—and, by extension, his larger relationship to the "new philosophy"—as merely tropical. As such, Empson's discussion of the separate planet prefigures the shift currently under way (as I have argued in this book) toward a literal rather than a strictly metaphorical analysis of the discourses of early modern science. Empson's discussions of Donne, in other words, invite readers to consider what it might mean that for Donne science was something more than simply a metaphor or a figure of speech.

Inspired in part by this example set by Empson, as well as by his care to embed autobiographical moments in his generally biographical discussion of "Donne the Space Man," and in part by a more general concern with method (in early modern science, in Donne's work, and in literary critical discourse), I would like to offer the following discussion of Donne and the new science. As I will argue, Donne renegotiates the crucial matters of embodiment and the contest between isolation and community through the strategic deployment of the autobiographical. And the particular mechanism that enables this renegotiation, I propose, is the discourse of exemplarity that was foundational to early modern science. While most other critical works that consider Donne's relation to the new philosophy, however, take his poems as their primary texts—a decision so thoroughly naturalized in the tradition of Donne criticism (including Empson's) as to seem inevitable—I want to offer an alternative. This alternative, as I will argue in the following pages, is not only important to any understanding of Donne's investment in the new philosophy, but is important precisely because it is his *only* text that can be said to be properly or strictly autobiographical: the *Devotions upon Emergent Occasions*.

---

[28] For a discussion of Empson's relationship to New Criticism, see (for example) Haffenden (2005).

II

Ingeniumque malum, numeroso stigmate, fassus, Pellitur
ad pectus, Morbique Suburbia, Morbus.

John Donne, *Devotions*
upon Emergent Occasions

Within a decade of Galileo's observations and revolutionary theoretical explanation of the existence of spots on the surface of the sun that was previously imagined as pristine—within a decade, that is, of the destruction of the very notion of *immaculate* nature—Donne was stricken with a near-fatal illness, one symptom of which was the appearance of spots on his body: "The Sicknes declares the infection and malignity thereof by spots."[29] Thus begins the thirteenth "station" of the *Devotions upon Emergent Occasions*, the 1624 meditation on the physiological and spiritual meanings of the unnamed disease (perhaps typhus or spotted fever) that nearly killed Donne in the preceding winter and that is chronicled in extraordinary detail in his book.[30] These mysterious "spots"—which not only bear witness to the severity of the disease but are indeed the very manifestations of the maculate nature of Nature—are for Donne doubly inscribed. On the one hand, the spots appear on his sick and perhaps dying body both as a sign of the progression of the disease and perhaps as harbingers of an impending abject hopelessness:

> In this *accident* that befalls mee now, that this sicknesse declares it selfe by *Spots*, to be a malignant and pestilentiall disease, if there be a *comfort* in the declaration, that therby the *Phisicians* see more cleerely what to doe, there may bee as much *discomfort* in this, That the malignitie may bee so great, as that all that they can doe, shall doe *nothing*; That an enemy *declares* himself, when he is able to subsist, and to pursue, and to atchive his ends, is no great comfort... O poore stepp toward being well, when these *spots* do only tell us, that we are worse, then we were sure of before. [*Devotions*, 67–8]

On the other hand, in the ensuing "Expostulation" (the title given to the second section of each of the text's twenty-three stations) in which the spots are understood as simultaneously physical *and* spiritual in nature ("*Lord*, if thou looke for a *spotlesnesse*, whom wilt thou looke upon?" [69]), the spots are ultimately interpreted to "belong" to the body of God's "Sonne," whose very incarnation itself is intended to enable him "to fetch" the spots/stigmata and "assume [them] to himselfe." Donne's Expostulation continues to think of the matter of the ownership of the spots: "When I open my *spotts*, I doe but present him with that which is *His*, and till I do so, I detaine & withhold *his right*... " (*Devotions*, 70).

This interpretive trajectory can be said to resolve the problem of the first kind of inscription through the mechanism of the second: the local problem of Donne's spots finds resolution in Christ's taking redemptive possession of Donne's natural spottedness. If the spots belong to the Son, then when Donne pauses to

---

[29] Donne (1987: 67). Subsequent references to this work appear parenthetically.
[30] Raspa discusses the matter of Donne's disease, which was likely one of two possibilities: typhus (or, more precisely, epidemic typhus) and the seven-day (or relapsing) fever; see Raspa (1987: xiii–ix).

contemplate them, when they become the objects of his meditation inscribed upon his flesh, they are, his theology assures him, no longer his. This means that they cannot be interpreted to signify meaningless death. Rather, the spots emerge through the practices of Donne's meditation (what he will call "confession") as the guarantors of salvation. Donne rounds this notion of the Son's salvational possession with a suitably apt celestial figuration:

> When therfore thou seest them upon me, as *His*, and seest them by this way of *Confession*, they shall not appear to me, as the *pinches of death*, to decline my feare to Hell; (*for thou hast not left thy holy one in Hell*, thy *Sonne* is not there) but these *spotts* upon my *Breast*, and upon my *Soule*, shal appeare to mee as the *Constellations* of the *Firmament*, to direct my Contemplation to that place, where thy Son is, thy *right hand*. [*Devotions*, 70]

But the primary and necessary act of seeing, which is in truth an act of reading, is God's, not Donne's and not the Son's. When God sees the spots on Donne's body and soul as the Son's, then Donne will be able to see them as signs of his salvation through the Son's act. Donne's reading of the spots completely depends upon God's prior act of reading. But what exactly is the nature of this act of reading? What text does God read and how does Donne account for its existence? This series of concerns culminates and concludes in the "Prayer" (the third and final section of each of the book's stations) in which Donne invokes the idea of human suffering as a form of God's instruction and offers thanks to God for allowing that he "can discerne thy *Mercie*, and find *comfort* in thy *corrections*." Noting "the *discomfort* that accompanies that phrase, *That the house is visited*, And that, *that thy markes, and thy tokens are upon the patient*," Donne nevertheless understands by them an even greater discomfort that mercifully has passed him over: that house "which is not *visited* by thee" and "what a *Wayve*, and *Stray* is that *Man*, that hath not thy *Markes* upon him." As for his own disease, Donne reaches a new understanding of the nature of its symptoms, both its fevers and its spots:

> These heates, *O Lord*, which thou hast broght upon this *body*, are but thy chafing of the *wax*, that thou mightest *seale* me to thee; These *spots* are but the *letters*, in which thou hast written thine owne *Name*, and conveyed thy selfe to mee; whether for a *present possession*, by taking me now, or for a future *reversion*, by glorifying thy selfe in my stay here, I limit not, I condition not, I choose not, I wish not, no more then the house, or land that passeth by any *Civill* conveyance. [*Devotions*, 70]

Inscription, then, leads Donne—through the figure of the body imagined as the softened wax necessary to bear God's impression and through the figure of the illness's spots understood as God's handwriting upon the diseased body—to the matter of embodiment. And indeed, this thirteenth station can serve as an epitome of Donne's *Devotions* more generally: presented as he is with the signs (or symptoms) of his physical disease, Donne dedicates his efforts to the discovery of their spiritual meaning. As an investigation into both the state of his natural body and at the same time the fate of his eternal soul, Donne's project seeks to comprehend the problem of embodiment. Donne realizes that the central mystery of life—manifest in its sheer variability, or the body's fragility, for instance, and in our inability to understand the physiology of the living organism—is inextricably linked with the central mystery of his faith, represented (or embodied) in Christ's incarnation.

The maculacy of human nature, visible in the many signs produced by or on the diseased body (the spots, the fever), naturally marks Donne's body for death. This was a matter of theological certitude. At the same time, and in fear of death and in fear for the fate of his soul, Donne wishes to interrogate his somatic maculacy and thus the spots enter into his meditative exegetical practice. If they are signs of God's corrective mercy (as Donne theorizes), then their first and most important meaning is that there *is* legible meaning: the spots (and symptoms) are not illegible—or, perhaps worse still, meaningless—but are in fact inscribed and therefore carry or convey meaning. As God's signs that seem to guarantee meaningfulness, the particular meaning the spots convey resides in that act by which Christ assumes Donne's spots (the signs of human sinfulness and fallenness) and the entire apparatus of Christian eschatological doctrine serves to reinvent the signs of the disease as signs that signify Christ's redemption of humankind. They can become the stars in the constellations that inscribe salvation in the figure of Christ. It is this argument that allows Donne to see the spots in a new way, as God's handwriting and God's "seale" that guarantees salvation. Donne's body bears the sign of God's ownership and Donne is willing to understand that this somatic sign of divine ownership extends to include his body as well as his soul: the healing of one represents the salvation of the other. The thirteenth station ends with a prayer in which Donne prays that God insure that "the closing of these bodily *Eyes* here, and the opening of the *Eyes* of my *Soule*, there, [are] all one *Act*" (*Devotions*, 70).

The discourse of the thirteenth Devotion makes clear that Donne's project—like the salvation of his eternal soul—depends upon the proper understanding of two fundamental and related issues: on the one hand, the nature of embodiment (or the body-soul relationship) and, on the other hand, the nature of God's method. In the fourth Expostulation, which marks the moment when the physician is summoned, Donne surveys scriptural discussions of physicians, medicines (in the form of simples), and the appropriate relationship the patient needs to establish with the physician: "Thou who sendest us for a blessing to the *Phisician*, doest not make it a curse to us, to go, when thou sendest" (*Devotions*, 22). Donne is clear that the curse—and the sin—would result from too dear and too exclusive an attachment to the physician at the expense of attention to God and his will. Citing Asa as an example to be repudiated of someone who "in his disease, *sought not to the lord, but to the Phisician*," Donne implores,

> Reveale therefore to me thy *method*, O Lord, & see, whether I have followed it; that thou mayest have glory, if I have, and I, pardon, if I have not, & helpe that I may.
> [*Devotions*, 22]

Set as it is within a moment in which Donne is thinking about physicians and medicines, the invocation here of "method" refers to the protocols associated with what we would today call medical diagnosis and the use of medical prescriptions to cure disease, a use of "method" that is consistent with a seventeenth-century definition of the term.[31] This is clearly the case at a later moment in the text when his physicians prescribe purgative medicines:

> The working of *purgative physicke*, is *violent* and contrary to *Nature*. O Lord, I decline not this *potion* of *confession*, how ever it may bee contrary to a *naturall man*. To take *physicke*, and *not according to the right method, is dangerous*. O *Lord*, I decline not that *method* in this *physicke*, in things that burthen my *conscience*, to make my *confession* to *him*, into whose hands thou hast put the *power* of *absolution*. [*Devotions*, 108]

At the same time, and more importantly, however, I would suggest that for Donne "method" carries with it those greater associations with our modern understanding of a defined and disciplined set of logical procedures, techniques, and practices dedicated to the realization of a particular objective. It is this sense in which Descartes will offer his *Discourse on Method* within a decade of the *Devotions*. (At the close of this chapter I will return to the Cartesian notion of method.) Donne's disease and his meditative text written in response to it together provide him the opportunity to construct an interpretation of God's method—of communication to the grievously ill Donne, of correction, of mercy, and of redemption. And, as I will argue further, in order to respond to this challenge, Donne must himself create a new method—of observation, of analysis, and of recording. In other words, Donne invents a discourse, which I will call *the new science writing*, as the apt vehicle for both the apprehension of God's method and a new articulation of a method that allows for the understanding of embodiment and materialism, together with a new model of human experience.

As will become clear immediately, this was an important but complicated relationship. Two passages from the *Devotions* illustrate at a glance the contrary movement in the text both toward and away from the new philosophy. Once Donne has passed the critical days of his illness and can leave his bed for the first time, he dilates upon the figure of what I am tempted to call his *Copernican* body:

> I am *up*, and I seeme to *stand*, and I goe *round*; and I am a *new Argument* of the *new Philosophie*, That the *Earth* moves round; why may I not beleeve, that the *whole earth* moves in a *round motion*, though that seeme to mee to *stand*, when as I seeme to *stand* to my *Company*, and yet am carried, in a giddy and *circular motion*, as I *stand*? [*Devotions*, 111]

This striking image is made possible by Donne's new science writing, which is itself underwritten (in part) by his intense awareness of the new philosophy created by Galilean astronomy. At the same time, however, Donne is by no means an apologist for the new philosophy. Indeed, he is often its satirist. The Copernican moment quoted above is all the more striking given its appearance within the more conventionally Ptolemaic setting of the *Devotions* in which Donne on many occasions displays something like a temperamental sympathy with the antiquated

---

[31] For an important discussion of method in early modern (philosophical) thought, see Gilbert (1960).

Ptolemaic universe that at times (in Meditation 10, for instance) seems to structure his imagination:

> This is *Natures nest of boxes*; The *Heavens* containe the *Earth*, the *Earth*, *Cities*, *Cities*, *Men*, And all these are *Concentrique*; the common *center* to them all, is *decay*, *ruine*; only that is *Eccentrique*, which was never made; only that place, or garment rather, which we can *imagine*, but not *demonstrate*, That light, which is the very emanation of the light of *God*, in which the *Saints* shall dwell, with which the *Saints* shall be appareld, only that bends not to this *Center*, to *Ruine*; that which was not made of *Nothing*, is not threatned with this annihilation. All other things are; even *Angels*, even our *soules*; they move upon the same *poles*, they bend to the same *Center*; and if they were not made immortall by *preservation*, their *Nature* could not keepe them from sinking to this *center*, *Annihilation*. [*Devotions*, 51]

These two passages together manifest what I have suggested throughout this discussion, that Donne's relationship to the new philosophy was clearly a conflicted one, especially on the thematic level in his works, in poetry or in prose, in which he is explicitly contemplating the new science or its new discoveries. But at the same time, Donne's works also exhibit a deeper engagement with the method of early modern science that is constructed upon the discourse of exemplarity. And in the case of the *Devotions*, it becomes especially clear the degree to which the discourse of exemplarity—upon which much of (early modern) science is predicated—is accessed by way of the autobiographical.

Although the status of Donne's *Devotions* as an autobiographical text has been much debated, for the purposes of this discussion I will assume the *Devotions* to participate in the widely varied textual and discursive practices of what will become identifiably autobiography precisely through the early modern period.[32] In identifying Donne's text—and project—as autobiographical, however, I am following the lead of a number of critics of seventeenth-century literature, including Janel Mueller whose essay on the *Devotions*, "The exegesis of experience," remains an important and influential contribution to our understanding of Donne's complex and challenging book.[33] Mueller argues that the *Devotions* (which, like *Essays in Divinity*, bears "a strangely potent blending of spiritual autobiography and textual analysis" [p. 4]) represents Donne's efforts toward "attaining [a] spiritual perspective on experience" (p. 6). To that end, she continues, Donne engages in "what may be called the exegesis of experience" (p. 7). Citing significant congruities between this work in the *Devotions* and a number of important sermons (especially the series of sermons devoted to exposition of Psalm 6),[34] Mueller concludes:

---

[32] See, for instance, Frost (1990, esp. Chs. 1 and 2).

[33] Mueller (1968).

[34] Psalm 6 is one of the so-called Penitential Psalms; it draws Donne's attention for a number of reasons, including its prayer for David's deliverance from not only from sin and his enemies, but also from sickness. Citing Evelyn Simpson's (conjectural) dating of the Psalm 6 series of sermons to the spring of 1623, Mueller notes that this would make these particular sermons "the latest extant known to have been preached before Donne's serious illness in the fall and winter of the same year, the illness that gave rise to the *Devotions*" (pp. 8–9).

Recognition of the accord in Donne's handling of Scriptural texts in the sermons on Psalm 6 and the *Devotions* should occasion no surprise, for they are essentially two versions (one public, one private) of an attempted running translation of a segment of experience into spiritual terms. It is more accurate, perhaps, to speak of a spiritual transliteration of experience, since Donne's ultimate preoccupation is always with particulars and how they influence and intimate the state of the soul. [p. 13][35]

And yet, one of the great complications in the *Devotions* arises precisely from the fact that "the state of the soul" (as Mueller has it) is never either wholly knowable nor entirely—that is, *exclusively*—central to the "anguished intellection" (p. 17) that readers such as Mueller have found evident throughout Donne's text. Indeed, the central challenge the book addresses lies in the indeterminacy of the nature of experience itself: never wholly material nor wholly immaterial, experience by its very nature resists the sort of transliteration into a fully spiritual register that would render it functional in Donne's exegetical practice and thereby instrumental to his larger devotional ambitions. The text bears many significant instances of this recognition of an ontological hybridity that characterizes what one might be tempted to call the human condition—at least as Donne conceives it. Though in Expostulation 22 (among other moments in the text) Donne will articulate a certain semiotic and assert that the body associates to the soul in the model of a signifier—"I know that in the state of my *body*, which is more *discernible*, than that of my soule, thou dost *effigiate* my *Soule* to me" (*Devotions*, 119)—this assertion does not seem sufficient to resolve the more fundamental problem that Donne cannot truly say what the body is and what the soul is, particularly in relation to the etiology of sin (and consequently of his physical sickness) that his meditations, expostulations, and prayers collectively seek to discover:

> My *God*, my *God*, what am I put to, when I am put to *consider*, and *put off*, the *root*, the *fuell*, the *occasion* of my *sicknesse*? What *Hypocrates*, what *Galen*, could shew mee that in my *body*? It lies deeper than so; it lies in my *soule*: And deeper than so; for we may wel consider the *body*, before the *soule* came, before *inanimation*, to bee *without sinne*; and the *soule* before it come to the *body*, before that *infection*, to be *without sinne*; *sinne* is the *root*, and the *fuell* of all *sicknesse*, and yet that which destroies *body* & *soule*, is in *neither*, but in *both together*; It is in the *union* of the *body* and *soule*; and, O my *God*, could I *prevent* that, or can I *dissolve* that? [*Devotions*, 118]

As this passage suggests, attempts to define sin—together with the related attempts to define body and soul—falter upon the prior problems of materialism in general, and embodiment in particular. In her recent book *John Donne, Body and Soul*, Ramie Targoff offers a sustained analysis of Donne's deep investment in the problem of embodiment across the many forms and genres in which he writes: letters, the songs and sonnets, the funeral poems, the divine poems, the sermons, and the *Devotions*.[36] Taking the relationship, in all its complexity, of the body and

---

[35] Mueller, later in her essay, will refer to "the instrumentality of figurative exposition to the spiritual translation of experience" (p. 17).

[36] Targoff (2008).

soul as "the defining bond" (p. 1) of Donne's life, Targoff traces Donne's engagement with the long theological and doctrinal debates about the nature of embodiment, as well as Donne's own and distinctive figurations and explorations of this defining issue:

> Donne was haunted throughout his life by feelings of the awkward dissociation between his body and soul: the tensions that arose between their respective needs; their irreconcilable states of health or illness; the occasional discrepancies between their objects of desire. At the same time, however, he was fully convinced of his body and soul's mutual dependence, and strove to create as harmonious and intimate a relationship between the two as possible . . . He was a dualist, but he was a dualist who rejected the hierarchy of the soul over the body, a dualist who longed above all for the union, not the separation, of his two parts. [p. 22]

If the intellection chronicled in the *Devotions* is anguished, it is so not only because thought (through written language, which is its sign) is called upon to express the burden of a translation of experience (from the somatics of disease to the certitude of spiritual redemption), but also because thought itself may be the expression of the defining incommensurability of body and soul.

As Elaine Scarry has argued, the *Devotions* constitutes a particularly compelling instance of the discourse of "volitional materialism."[37] The desire to understand materialism as voluntary (an ambition modeled on God's example, including his willful embodiment as human) marks for Scarry a certain set of mental and authorial habits that characterize Donne's poetry and prose. While Scarry acknowledges this desire is not exclusive to Donne but in fact characterizes the early modern period in general and, in particular, "the rather glacial change from a religious to a secular and scientific world," the finer details of this glacial change have not yet been articulated. I would like to attempt to sketch the broad outlines of this history by arguing that, more than mere mental habit, Donne's volitional materialism is itself organized by the emergent culture of early modern science.[38] It is the resulting system of artifaction (as I will call it), forged in this instance by the juxtaposition of the clinical case study with the spiritual memoir, that serves to construct the (autobiographical) self as both the normative and ideal artifact of discourse.

I would like to argue that, more than mere mental habit, Donne's confrontation with materialism and his attempts to renegotiate its relation to both body and "self" are themselves organized by an epistemology in the process of forging the means "to convert" (recalling Elizabeth Spiller's words) "accounts of personal experience into new stories of universal truth."[39] It is this epistemology that in first-wave science and literature criticism was uniformly construed as the proprietary domain of science that produced knowledge of the natural world; this knowledge—along with the idea of science as a master discourse—was then reflected in literary works said to be constructed (at least in part) under the influence of science. In

---

[37] Scarry (1988: 71).
[38] For discussions of the *Devotions* and early modern medical science, see Pender (2003) and Kuchar (2005, esp. 151–79).
[39] Spiller (2004: 15).

second-wave criticism (or what I have also called the discursive model), this epistemology is more appropriately understood as shared equally between emergent science and literary culture. Indeed, this epistemology can best be understood as dispersed across the culture, more generally. Such an understanding locates both science and literature within a mutually sustaining network that denies privilege or priority to either one. In this way, the notion of influence—especially when it is imagined as unidirectional—becomes a meaningless category. As such, with the scientific and the literary as manifestations (or articulations) of an underlying epistemology, we can better appreciate the symmetrical relationship between them.

While Donne's relationship to the new science was clearly a complex and perhaps troubled one and speaks to a more or less individual set of sympathies and antipathies, the argument has been made that the advent of the new science within the sixteenth and seventeenth centuries had a deleterious effect on the traditions of spiritual autobiography which, as Kate Gartner Frost has argued, were characterized by "the use of typology, fictive devices, and deliberate structure":

> Chief among [the intellectual and scientific changes impacting spiritual autobiography] was the so-called Copernican revolution: No longer could the individual place himself within a coherent moral and spiritual structure; no longer could he see his life played out on a recognizable stratum of the universal hierarchy; no longer was one's very physical existence reflected in and magnified by the speculum of correspondences that composed the late-medieval cosmos.[40]

Frost also suggests that alongside "new" philosophy as a determining force in the reconceptualization of spiritual autobiography there are other, equally powerful, social and discursive forces that contributed to these transformations that, taken together, "inexorably moved the writer's focus toward the subjective" (p. 78).[41] There is much value in Frost's argument, especially if we understand the transformation of spiritual autobiography into autobiography as at least in part characterized by this move toward the subjective—and that the subjective itself becomes more fully available (even if in part compensatorily) in the absence of earlier (that is, medieval and scholastic) schemas rendered obsolete by the advent of the culture of science. At the same time, however, there are ways in which autobiographical writing such as the *Devotions* is enabled more positively precisely by the protocols and practices of that culture of science. The *Devotions* is an especially compelling text on this score because in it we can trace the contour lines of confrontation between the older, essentially scholastic, systems of correspondences that determined meaning in the world and the newer, essentially scientific system of induction and experiment that created meaning in the world.

I would like to focus attention on one particular aspect of this emergent culture of science and the ways in which we can see its important effects in the *Devotions*: first, the ways in which experience becomes understood and validated as evidential,

---

[40] Frost (1990: 78).

[41] So radical were these changes, Frost argues, that "critics today throw doubt on the existence of 'real' autobiography" prior to the seventeenth or eighteenth century (p. 78).

and second, the ways in which the evidential is deployed in the production of meaning that is itself understood to be universal. When indexed to a text that is autobiographical, this becomes a matter of a different form of transliteration: from the personal to the generalized. The opening line of the *Devotions* bears the weight of this renegotiation of experience, together with the desire toward universalizing that is one of the hallmarks of both Donne's devotional text and emergent science:

> Variable, and therefore miserable condition of Man; this minute I was well, and am ill, this minute. [*Devotions*, 7]

The apparent clarity of this desire to create universally true knowledge—that the fallen, and therefore "natural" condition of "man" is misery—itself obscures a very complex set of logical and inductive maneuvers which I want to call scientific. What is it that allows Donne to move from "Man" in the opening phrase, to "I" in the second? What allows for *this* sort of transliteration? There are two answers to these questions. The first corresponds to what I earlier called the scholastic epistemological model of correspondences that dictated a certain, and predicted, meaning to such a sentence: "Man's condition in the world is miserable; I am a man, therefore I am miserable." The logic of this is undeniable. And yet, this is not the only—nor the presiding—logic within the *Devotions*. The second answer to our question points to a more rigorously inductive response that will come to typify the scientific epistemological mode more generally: "I am miserable; I am a man, therefore the condition of man in the world is miserable." In the former statement, which is wholly teleological in form and nature, that which the statement would prove is itself accepted a priori as a truth: "Man's condition is miserable." In this sense, this statement cannot be said to prove anything at all, other than its own perfect solipsism, and as such it functions in the opening moment of the *Devotions* as a statement of faith or belief. For the latter, to the contrary, it is Donne's ontology— he is a man and he is miserable—that are taken as a priori truths and the conclusion, the thing that is proved, is the universally miserable condition of all "men."

But what is it, we may ask, that allows Donne's argument to proceed along the trajectory outlined in this second model? The answer, I would argue, is a particular understanding of the experience of embodied experience. Donne is empowered to conclude that the condition of man is miserable not because this is his starting point, but rather because this is his conclusion, and it is a conclusion that is not based on faith or belief, but rather on first-hand experience that establishes the validity of his argument.

The nature of experience, then, is central to the work of the *Devotions*, even as the creation and communication of experience becomes foundational to the practices of science. In *Discipline and Experience*, Peter Dear identifies the crucial distinction between Aristotelian definitions of experience (such as presided over medieval and scholastic epistemology) and the "scientific" understanding of experience emergent in Europe in the seventeenth century as fundamentally a matter of the importance of the particular—and particularized—historical event (or, in its more fully evolved form, the "event experiment"):

Throughout the seventeenth century, the touchstone for definitions of experience in the literate philosophical discourse of Western Christendom remained the writings of Aristotle. An "experience" in the Aristotelian sense was a statement of *how things happen* in nature, rather than a statement of *how something had happened* on a particular occasion: the physical world was a concatenation of established but sometimes wayward rules, not a logically integrated puzzle. But the experimental performance, the kind of experience upheld as the norm in modern scientific practice, is unlike its Aristotelian counterpart; it is usually sanctioned by reports of historically specific events.[42]

Further, Dear argues that the revolutionizing power of science lies in the "metaphorical functioning of experiment" (p. 158)—the particular mechanism that allows one to argue that the occurrences within any given particular historical event/experiment are generalizable to the greater world. Citing the work of Pascal with Torricellian tubes and William Gilbert's with the *terrella* (the "little earth," a machined spherical lodestone), Dear demonstrates how early modern science breaks down and revises the traditional art–nature distinction in such as way as to render experimental (that is, contrived or *artificial*) results not so much semiotically metaphorical, as rather *mimetically* metaphorical. For Pascal, then, the Torricellian tube that demonstrates the possibility of a vacuum "does not here signify an aspect of the world," but rather is "actually *like* that aspect of the world in some essential way—there is a movement toward identity." Similarly, for Gilbert, the *terrella* demonstrates the magnetic characteristics of the earth:

> The little earth and the great earth are interactively related so as to set up a kind of identity: the *terrella* does not simply imitate the earth, so as to elucidate properties of the latter, and the earth does not represent a giant *terrella*, so as to elucidate properties of magnets. Instead, the earth *is* a magnet, and the *terrella*, possessing the proper shape for a magnet, *is* a little earth. [p. 159]

In similar fashion, Donne is able to establish the generalized validity of his conclusions—which are not limited, of course, to statements about the misery of "man," but are more centrally about the discovery of salvation through the various processes of disease and recovery. The conclusions he reaches through the *Devotions* emerge as the consequences of his experience of disease, rather than as simply functions of a certain set of religious beliefs. It is for this reason, I believe, that the "lessons" learned through the narrative offered in the *Devotions* are anything but surprises. Donne concludes that despite all of his exegetical and interpretive powers, and despite his vast learning and powers of intellection, in the end he can neither heal his own body, nor save his own soul from damnation:

> To cure the *sharpe accidents* of *diseases*, is a great worke; to cure the *disease it selfe*, is a greater; but to cure the *body*, the *root*, the *occasion* of *diseases*, is a worke reserved for the great *Physitian*, which he doth never any other way, but by *glorifying* these *bodies* in the next world. [*Devotions*, 117–18]

---

[42] Dear (1995: 4). See also Chs. 3 (on "experimental events") and 5 ("The uses of experiment").

The difference, though, between these lines as conclusions drawn from *something that had happened* because of a disease, and similar lines (or, indeed, even these same lines) as predictions of *what happens* because of a disease, is absolute. It is the same distinction that obtains between the scientific narrative of what happened to *my* body in illness and an Aristotelian notion of what happens to the (abstract) body in illness.

At this point we may well ask what in particular renders Donne's *Devotions* distinct from any other prior or contemporary personal accounts of an event, since I would by no means want to suggest that all such accounts are either necessarily autobiographical or properly scientific; or, to think of this in another way, what makes Donne's experience of his nearly fatal illness an experiment? There are certainly a number of necessary conditions: Donne can write his account only after the fact of his illness and recovery (it is not, after all, a journal or diary); writing an account of this experience is only possible once Donne has re-created both his experience and himself as artifacts of interpretive discourse; and the entire project is underwritten by an operational identity of the observed and observing self. The issue before us becomes, then, one of understanding the method of Donne's transformation of his experience (in all of its complexity) into something of universal significance and meaning. In the face of these concerns, what can we say about the poetics of exemplarity?[43]

To address these crucial questions, I would like to consider the three-station portion of the *Devotions* that constitutes what I take to be the pivotal moment in the sequence, the sixteenth through the eighteenth station in which Donne famously contemplates the tolling of the bells. This sequence is significant for two particular reasons. On the one hand, these three linked Meditations collectively serve an important narrative function as they mark what will prove to be the turning point in the course of the disease. The immediately preceding Meditations marked the gravest extent of the progression of the disease: the thirteenth and fourteenth Meditations take up the sudden and distressing appearance of spots on Donne's body and the fifteenth marks the moment when Donne is no longer able to sleep. The nineteenth Meditation marks Donne's emergence from the "criticall dayes" (*Devotions*, 71) of the worst of what the disease will do and announces the hopefulness of eventual recovery: "At last, the Physitians, after a long and stormie voyage, see land; They have so good signes of the concoction of the disease, as that they may safely proceed to purge" (*Devotions*, 97). On the other hand, this sequence of three Meditations also constitutes that moment in which Donne is able to negotiate the elevation of his personal experience to the level of universal meaning. While the tolling of the bells announces critical stages in Donne's narrative of disease and faith, they perhaps more importantly can be said to structure the theory of exemplarity that insures the success of Donne's entire project.

---

[43] It is worth noting here that the following discussion of exemplarity arising from the necessarily autobiographical nature of the embodied self's experience of embodiment differs markedly (and perhaps entirely) from the ongoing practices of empirical science that (on one account, at least) moves inexorably in the direction of what is commonly called objectivity: the carefully disciplined absence (which is, of course, wholly rhetorical and performative in nature) of the acting subject.

The sixteenth Meditation is notable for a number of reasons, including the fact that it represents the first time in the book that the exterior world—quite literally, the world outside Donne's sick room—intervenes. Indeed, the Meditations to this point are marked by a profound insularity that is itself a function of the perfectly isolating effects of the disease. Titles to the first fifteen Meditations suggest the power of the disease to separate Donne and isolate him and his body from friends and society in general:

1: *The first alteration, The first grudging of the sicknesse.*
2: *The strength, and the function of the Senses, & other faculties change and faile.*
3: *The Patient takes his bed.*
4: *The Physician is sent for.*
5: *The Phisician comes.*
6: *The Phisician is afraid.*
7: *The Phisician desires to have others joyned with him.*
8: *The King sends his owne Phisician.*
9: *Upon their Consulation, they prescribe.*
10: *They find the Disease to steale on insensibly, and endeavor to meet with it so.*
11: *They use Cordials, to keep the venim and Malignitie of the disease from the Heart.*
12: *They apply Pidgeons, to draw the vapors from the Head.*
13: *The Sicknes declares the infection and malignity thereof by spots.*
14: *The Phisicians observe these accidents to have fallen upon the criticall dayes.*
15: *I sleepe not day nor night.*

"As Sicknesse is the greatest misery," Donne reasons in the fifth Meditation, "so the greatest misery of sicknes is *solitude*" (*Devotions*, 24). And this observation inaugurates a sustained consideration of the fact of isolation that results from a general fear of contagion ("the infectiousnes of the disease deterrs them who should assist, from coming; Even the *Phisician* dares scarse come" [*Devotions*, 24–5]) and the corresponding emotional impact of isolation experienced as solitude. "*Solitude*," Donne writes, "is a torment, which is not threatned in *hell* it selfe" (*Devotions*, 25).

Donne knows perfectly well that the experience of solitude is, in a manner of speaking, a function of his body in disease as understood—or perhaps misunderstood—by others:

When I am dead, & my body might infect, they have a remedy, they may bury me; but when I am but sick, and might infect, they have no remedy, but their absence and my solitude. It is an *excuse* to them that are *great*, and pretend, & yet are loth to come; it is an *inhibition* to those who would truly come, because they may be made instruments, and pestiducts, to the infection of others, by their comming. And it is an *Outlawry*, an *Excommunication* upon the *patient*, and seperats him from all offices not onely of *Civilitie*, but of *working Charitie*. A long sicknesse will weary friend at last, but a pestilentiall sicknes averts them from the beginning. [*Devotions*, 25]

But Donne also knows that solitude as such is contrary not only to nature, but to God's nature, as well. "Meere *vacuitie*, the first *Agent*, *God*, the first *instrument* of *God*, *Nature*, will not admit; Nothing can be utterly *emptie*; but so neere a degree towards *Vacuitie*, as *Solitude*, to bee but one, they love not" (*Devotions*, 25). As he continues it becomes clear that isolation's production of solitude is indeed contrary to the desire of both man and God for communion: "*God* himself," Donne avers, "wold admit a *figure* of *Society*, as there is a plurality of persons in *God*, though there bee but one *God*; & all his externall actions testifie a love of *Societie*, and *communion*" (*Devotions*, 25). All things tend naturally toward communion: "*Orders of Angels*" and the "*many mansions*" in heaven; "in *Earth*, *Families*, *Cities*, *Churches*, *Colleges*, all *plurall things*;" God "saw that it was not good, for man to bee *alone*, therefore *hee made him a helper*; and one that should helpe him so, as to increase the *number*, and give him *her owne*, and *more societie*." Given, then, that "for things of this world, their blessing was, *Encrease*," Donne believes he need not "aske leave to think, that there is no *Phenix*; nothing singular, nothing alone" (*Devotions*, 25). With the accumulated weight of "encrease" and "a *pluralitie* in every *Species* in the world," Donne concludes confidently that "the abhorrers of *Solitude*, are not solitary; for *God*, and *Nature*, and *Reason* concurre against it" (*Devotions*, 26). The concluding sentence of the fifth Meditation returns to the figure of the body and poses the complicated question of the relation of the body and the soul to one another. Donne's "infectious bed" is, he comes to understand, "equall, nay worse then a *grave*, that thogh in both I be equally alone, in my bed I *know* it, and *feele* it, and shall not in my *grave*: and this too, that in my bedd, my soule is still in an infectious body, and shall not in my grave bee so" (*Devotions*, 26).

Part of the challenge that Donne faces in his sickness and in his account of it is to effect a movement from solitude toward communion—or, in another vocabulary, from the individual to the universal. This process begins in earnest (as suggested above) in the sixteenth Meditation with the tolling of the bells, the first harbinger of the persistence of a meaningful—because signifying—world outside of Donne's suffering body: "*From the bels of the church adjoyning, I am daily remembred of my buriall in the funeralls of others*" (*Devotions*, 81). Donne begins this Meditation with a reference to Gerolamo Maggi, "a *Convenient Author*, who writ a *Discourse of Bells* when hee was Prisoner in *Turky*," and includes discussion of the Turks, having conquered Constantinople, melting church bells for the production of ordnance to be used in the defense of the city. Donne notes, however, how much greater Maggi's work would have been had he "beene my *fellow Prisoner* in this *sicke bed*" (*Devotions*, 81–2).[44] This is due to both incapacity brought by disease and to the tolling of funeral bells Donne can hear from his sick bed: "I have heard both *Bells* and *Odrnance*," Donne confesses, "but never been so much affected with those, as with these *Bells*" (*Devotions*, 82). Like ordnance, the bells constitute something of an assault, intruding upon Donne's solitude; but, as such the bells are also eruptions of a reminder of community that exists beyond the confines of the sick room and that includes the sick man: "Here the *Bells* can scarse solemnise the funerall of any person, but that I knew him, or knew that hee was my *Neighbour*: we dwelt in houses neere to one another before, but now hee is gone

---

[44] Gerolamo Maggi (or Hieronymus Magius) (1608).

into that house, into which I must follow him" (*Devotions*, 82). Donne then considers experience by proxy: the children of the wealthy whose punishment is dealt to surrogate children ("other *Children* are corrected in their *behalfe*, and in their *names*, and this workes upon them, who indeed had more deserved it"); or spectators at a public execution who "if they would aske, for what dies that Man, should heare their owne faults condemned, and see themselves executed, by *Atturney*"; or when any of us sees someone preferred and we "thinke our selves, that wee might very well have beene that *Man*." Donne draws the lesson clearly:

> Could I fit my selfe, to *stand*, or *sit* in any Mans *place*, & not to lie in any mans *grave*? I may lacke much of the *good parts* of the meanest, but I lacke nothing of the *mortality* of the weakest; They may have acquired better *abilities* than I, but I was borne to as many *infirmities* as they. To be an *incumbent* by lying down in a *grave*, to be a *Doctor* by teaching *Mortification* by *Example*, by *dying*, though I may have *seniors*, others may be *elder* than I, yet I have proceeded in a good *University*, and gone a great way in a little time, by the furtherance of a vehement *fever*; and whomsoever these *Bells* bring to the ground to day, if hee and I had beene compared yesterday, perchance I should have been thought likelier to come to this preferment, then, than he. [*Devotions*, 82–3]

Exemplarity begins to emerge in this Meditation as crucial to Donne's interpretive and devotional project. "A man extends to his *Act*," he writes, "and to his *example*; to that which he *does*, and that which he *teaches*." This applies to the bells, as well, and the ringing of the funeral bells for another man serves "to bring him to mee in the application" (*Devotions*, 84).[45] The bells, then, carry multiple identifications: eruptions of the external world into Donne's sick room, the calling together of Christian believers ("the *evill spirit* is vehemently vexed in their ringing, therefore, because that action brings the *Congregation* together, and unites *God* and his *people*" [*Devotions*, 83]); and, most significantly, the voice of the dead:

> I make account that I heare this dead brother of ours, who is now carried out to his *buriall*, to speake to mee, and to *preach* my *funerall Sermon*, in the voice of these *Bells*. In him, O God, thou hast accomplished to mee, even the request of *Dives* to *Abraham*; *Thou hast sent one from the dead to speake unto mee*. He speaks to mee aloud from that *steeple*; hee whispers to mee at these *Curtaines*, and hee speaks thy words; *Blessed are the dead which die in the lord . . .* [*Devotions*, 85]

In the seventeenth Meditation ("*Now, this Bell tolling softly for another, saies to me, Thou must die*" [*Devotions*, 86]), Donne further refines his meditation on the figure

---

[45] In Exposulation 16, Donne offers an interesting (and unexpected) consideration what he calls "*historicall pictures*," but which are better known as religious icons:

> We cannot, wee cannot, O my *God*, take in too many *helps* for religious *duties*; I know I cannot have any better *Image* of *thee*, than thy *Sonne*, nor any better *Image* of *him*, than his *Gospell*: yet must not I, with thanks confesse to thee, that some *historicall pictures* of his, have sometimes put mee upon better *Meditations* than otherwise I should have fallen upon? [*Devotions*, 84]

Donne attributes the efficacy of "*historicall pictures*" to a manifestation of God's restraint: "in making us *Christians*, thou diddest not destroy that which we were before, *naturall men*, so in the exalting of our religious devotions now we are *Christians*, thou hast beene pleased to continue to us those *assistances* which did worke upon the affections of *naturall men* before . . ." (*Devotions*, 84–5).

of the voice of the dead carried in the tolling of the bells. His first move here, following (as it does) upon the logic of the sixteenth Meditation, is to confront the realization that the relationship between the dead man whose funeral bells he hears and the sick (and perhaps sick to death) Donne may not end with the dead man—however vocal he may be in the ringing of the bells—serving as proxy. "Perchance hee for whom this *Bell* tolls, may bee so ill," Donne muses, "as that he knows not it *tolls* for him," and it is at this point that proxy becomes identity: "And perchance I may thinke my selfe so much better than I am, as that they who are about mee, and see my state, may have caused it to toll for mee, and I know not that" (*Devotions*, 86). This transformation causes, in turn, Donne's further realiza-tion—that the "*Catholike, universall*" nature of the church is underwritten by an absolute community of believers ("All *mankinde* is of one *Author*") that serves to unite all into one body: "When [the church] *baptizes a child*, that action concernes mee; for that child is thereby connected to that *Head* which is my *Head* too, and engraffed into that *body*, whereof I am a member." The conclusion rings, we might say, with a certain inevitability: "The *Bell* doth toll for him that *thinkes* it doth" (*Devotions*, 86), a formulation that stands (for us) as a kind of prelude to the most famous of passages from the *Devotions*:

> who bends not his *eare* to any *bell*, which upon any occasion rings? but who can remove it from that *bell*, which is passing a *peece of himselfe* out of this *world*? No Man is an *Iland*, intire of it selfe; every man is a peece of the *Continent*, a part of the *maine*; if a *Clod* bee washed away by the *Sea*, *Europe* is the lesse, as well as if a *Promontorie* were, as well as if a *Mannor* of thy *friends*, or of *thine owne* were; Any Mans *death* diminishes *me*, because I am involved in *Mankinde*; And therefore never send to know for whom the bell tolls; It tolls for *thee*. [*Devotions*, 87]

Donne reasons that this expansion of the very nature of the bells—as a sign of a person's death and funeral rituals, as the voice of that dead person speaking out to Donne (and the rest of us), as a figuration of our shared and mutual fates—is itself a function of God's will; it is one of God's ways of "*drawing light out of darknesse*" (*Devotions*, 87) and "*to raise strength out of weaknesse*" (*Devotions*, 88). The sound of the tolling of the bell is, Donne concludes, the voice of God: "Thy *voice*, thy *hand* is in this *sound*" (*Devotions*, 88). The tolling of the bells continues, as it were, this evolution toward divinity. The seventeenth Prayer continues to trace this spiritual movement toward heaven, toward the voice of God:

> O eternall and most gracious *God*, who hast beene pleased to *speake* to us, not onely in the *voice of Nature*, who speakes in our *hearts*, and of thy *word*, which speakes to our *eares*, but in the speech of *speechlesse Creatures*, in *Balaams Asse*, in the speech of *unbeleeving men*, in the confession of *Pilate*, in the speech of the *Devill* himselfe, in the *recognition* and *attestation* of thy *Sonne*, I humbly accept thy *voice*, in the sound of this sad and funerall *bell*. [*Devotions*, 89]

This attention to the sound of God's voice in the bells leads Donne to an astonishing moment in which he adds his own voice to what he characterizes as "this *whole Consort*" (*Devotions*, 88) of divine (or divinely inspired voices) and speaks to God in the voice of David and in the voice of Christ:

> As *death is the wages of sinne*, it is *due* to mee; As death is *the end of sicknesse*, it belongs to *mee*; And though so disobedient a *servant* as I, may be afraid to *die*, yet so mercifull a *Master* as thou, I cannot be afraid to *come*; And therefore, *into thy hand*, O my *God*, *I commend my spirit* . . . [*Devotions*, 89]

As this passage would seem to suggest, Donne's spiritual (and theological) project is virtually completed at this point. The work that remains is achieved in the eighteenth Meditation ("*The bell rings out, and tells me in him, that I am dead*" [*Devotions*, 91]) and depends upon the final (re)interpretation of the tolling of the bells.

The eighteenth Meditation considers at its outset two matters: the departure of the soul upon death, and the state or nature of what remains once the soul has departed from the body, as has happened in the case of the man whose funeral bells have so powerfully attracted, or perhaps demanded, Donne's attention. At first mention of the departing soul, Donne makes clear that clarity of knowledge is not to be hoped: "His *soule* is gone; *whither*? Who saw it *come in*, or who saw it *goe out*? No *body*; yet every body is sure, he *had one*, and *hath none*" (*Devotions*, 91). When Donne imagines asking those who might be imagined to know answers to these questions, he finds philosophers and "*Philosophicall Divines*" (*Devotions*, 91) both hopelessly fragmented into diverse—and antagonistic—responses, even within their own disciplines. Asked about the origin of the soul, philosophers are likely to say that the soul is nothing but "*temperament*" or the "*just and equall composition of the Elements in the body*" (*Devotions*, 91). The "*Philosophicall Divines*," for their part, are likely to argue that the soul is a "*separate substance*" that enters our bodies "by *generation & procreation* from *parents*," others, that the soul is placed into man by "*immediate infusion from God*" (*Devotions*, 91). And when confronted by the perhaps harder-still question of "what becomes of the *soules* of the *righteous*, at the *departing* thereof from the *body*" (*Devotions*, 91–2), the confusion persists: some will argue "*they attend an expiation, a purification in a place of torment*," others, that they merely wait "*in a place of rest; but yet, but of expectation*," and still others, that they "*passe to an immediate possession of the presence of God*" (*Devotions*, 92). Donne resolves this dilemma for himself by following the example of St. Augustine, "who satisfies himselfe with this: *Let the departure of my soule to salvation be evident to my faith, and I care the lesse, how darke the entrance of my soule, into my body, bee to my reason. It is the going out, more than the coming in, that concernes us*" (*Devotions*, 92).[46]

---

[46] Citing Donne's wish (expressed in Expostulation 1) for a homological relationship between body and soul ("why is there not alwayes a *pulse* in my *Soule*, to beat at the approch of a tentation to sinne?" [*Devotions*, 8]), Targoff offers this interesting—and, I believe, accurate—formulation:

> Whereas the body intuits its own illness even before the illness has manifested itself in visible symptoms, the soul repeatedly fails to detect its own sins. The resonant suggestion that the soul should have a pulse conveys the extent of Donne's somatic imagination: the ideal soul is a corporeal soul, with its own heart beating. [p. 140]

And Donne appears to be good to his word: his concern in the *Devotions* (and this is true for his poetry, as well, I would suggest) lies not with the matter of the placing of the soul in the body, but the consequences arising upon its departure out of the body.[47] But what Donne's declared reliance on St. Augustine does not indicate is that his attention, while concerned overwhelmingly to "the going out," has a double focus: on the movement of the soul toward heaven (in whatever fashion that motion is imagined, either immediate or mediated in some fashion) and on the body that remains behind. Although Donne cannot answer the question he asks of the soul of the man whose bells he has heard and whither it went, he generates a benevolent answer out of his own goodwill, together with a small share of self-interest: "This *soule*, this *Bell* tells me is *gone out*; *Whither?* Who shall tell me that? I know not *who* it is; much lesse *what* he was; The condition of the Man, and the course of his life, which should tell mee *whither* hee is gone, I know not" (*Devotions*, 92). Donne cannot tell himself, nor can he ask any who knew the man, his way in the world or his passing. But he can consult with his own "*Charity*":

> I aske that; & that tels me, *He is gone to everlasting rest*, and *joy*, and *glory*; I owe him a good *opinion*; it is but *thankfull charity* in mee, because I received *benefit* and *instruction* from him when his *Bell* told. [*Devotions*, 92]

For Donne, this departed soul has already both reached its salvation and fulfilled its instructive purpose for attentive survivors—and listeners. But what is especially striking at this point is Donne's turn of attention to the now dis-inhabited body. "But for the *body*," Donne admits and laments, "How poore a wretched thing is *that?* wee cannot express it *so fast*, as it growes *worse* and *worse*" (*Devotions*, 92). Donne then reads a eulogy on the dead body itself:

> That *body* which scarce *three minutes* since was such a *house*, as that that *soule*, which made but one step from thence to *Heaven*, was scarse thorowly content, to leave that for *Heaven*: that *body* hath lost the *name* of a *dwelling house*, because none dwels in it, and is making haste to lose the name of a *body*, and dissolve to *putrefaction*. Who would not bee affected, to see a cleere & sweet *River* in the *Morning*, grow a *kenell* of muddy land water by *noone*, and condemned to the saltnesse of the *Sea* by *night*? And how lame a *Picture*, how faint a *representation*, is that, of the precipitation of mans body to *dissolution? Now* all the parts built up, and knit by a lovely *soule*, now but a *statue* of *clay*, and *now*, these limbs melted off as if that *clay* were but *snow*; and now, the whole *house* is but a *handfull of sand*, so much *dust*, and but a *pecke of Rubbidge*, so much bone. [*Devotions*, 92–3]

For Donne, not only is this the fantasized and dreaded end of our mortal bodies on earth, it is also *telos* figured in birth: not our births from our "*naturall Mother*," but rather the birth from our "*Mother in law . . .* the *Earth*" out of our graves: "In the

---

[47] Targoff discusses the competing theories meant to account for the creation of the soul and the way in which it is made to abide in the body: infusionism (direct infusion into the body by God) or propagationalism (the generation of the soul within the body), as well as the theological challenges each represented (e.g. how the soul might contract Original Sin, for instance). Also, for a discussion of Donne and his "unusually active relationship to his mortality" (p. 217), see Targoff (2006).

wombe of the Earth, wee *diminish*, and when shee is *delivered* of us, our *grave* opened for another, wee are not *transplanted*, but *transported*, our *dust* blowne away with *prophane dust*, with every wind" (*Devotions*, 93).[48]

The crisis represented by this profound concern for the dissolution of the dead body and the vexing matter of its relation to our souls (especially after death) have for Donne only one solution.[49] The Prayer that completes the eighteenth Meditation contains what is perhaps the most important argument to be found anywhere in the *Devotions* and it has everything to do with the forging of a new understanding of embodiment. Donne is clearly not willing to part with the departed body and his solution to this problem is twofold. On the one hand (as has been noted by many readers and critics) Donne's heaven at the end of time is populated by souls reunited with their exact, though purified, material bodies—what Donne here calls "the full *consummation* of all, in *body* and *soule*," even as he prays, too, that the Son should "have *societie* of humane *bodies* in *heaven*" (*Devotions*, 96).[50] On the

---

[48] There are many other examples in the text in which Donne characterizes the fallen state of the natural body; see, for instance, Meditation 22, which includes the following passage:

> When therefore I tooke this *farme*, undertooke this body, I undertooke to *draine*, not a *marish*, but a *moat*, where there was, not water *mingled* to offend, but all was *water*; I undertooke to *perfume dung*, where no one part, but all was equally *unsavory*; I undertooke to make such a thing *wholsome*, as was not *poison* by any manifest quality, *intense heat*, or *cold*, but *poison* in the *whole substance*, and in the *specifique forme* of it. [*Devotions*, 117]

[49] This problem may also have one significant danger; Expostulation 18 offers a cautionary explanation of the (anthropological) for the emergence of Idolatry:

> I satisfie my selfe with this; that in those *times*, the *Gentiles* were overfull, of an over-reverent respect to the *memory of the dead*: a great part of the *Idolatry* of the *Nations*, flowed from that; an over-amorous devotion, an *over-zealous celebrating*, and *over-studious preserving* of the *memories*, and the *Pictures* of some *dead persons*: And by *the vaine glory of men, they entred into the world*; and their *statues*, and *pictures* contracted an opinion of *divinitie*, by *age*: that which was at first, but a *picture* of a *friend*, grew a *God* in time. [*Devotions*, 93–4]

(The solution to *this* problem, Donne suggests, lies in the very nature of Christ, "the *first begotten of the dead*," who is "my *Master* in this *science* of *death*" (*Devotions*, 94) as exemplum.)

This passage is especially interesting if one contemplates Donne's own final act of contemplation, as reported by Walton in his account of Donne's life and his final illness and death (Walton 1927, rpt. 1950):

> A Monument being resolved upon, Dr. Donne sent for a Carver to make for him in wood the figure of an Urn, giving him directions for the compass and height of it; and to bring with it a board of the just height of his body. These being got: then without delay a choice Painter was got to be in a readiness to draw his Picture, which was taken as followeth. Several Charcole-fires being first made in his large Study, he brought with him into that pace his winding-sheet in his hand, and, having put off all his cloaths, had this sheet put on him, and so tyed with knots at his head and feet, and his hands so placed, as dead bodies are usually fitted to be shroweded and put into their Coffin, or grave. Upon this Urn he thus stood with his eyes shut, and with so much of the sheet turned aside as might shew his lean, pale, and death-like face, which was purposely turned toward the East, from whence he expected the second coming of his and our Saviour Jesus. [p. 78]

See also the account offered in Bald (1970: 525–30).

[50] There is a similar moment in Donne's final sermon, "Death's Duel," preached before Charles, 25 February 1631 in Donne (1990). After considering the fate of human remains (worms and the dispersion of dust), Donne assures his auditors of God's infinite ability:

> This death of incineration and dispersion is, to natural reason, the most irrecoverable death of all; and yet, *Domini Domini sunt exitus mortis, unto God the Lord belong the issues of death*; and by recompacting this dust into the same body, and reanimating the same body with the same soul, he

other hand—and perhaps more radically—Donne's resolution to the dead body is to identify it as inseparable from the living body. Or rather, the living body is understood as always dying, always already dead:

> *I am dead*, I was *borne dead*, and from the first laying of these *mud-walls* in my *conception*, they have *moldred* away, and the whole course of *life* is but an *active death*. Whether this *voice* [figured in the tolling bells] *instruct* mee, that I am a *dead man now*, or *remember* me, that I have been a *dead man* all this while, I humbly thanke thee for speaking in this *voice* to my *soule*. [*Devotions*, 96]

The eighteenth Meditation constitutes the culminating moment in the *Devotions*. This is true in terms of the narrative of Donne's sickness: the crisis of the disease is negotiated through this point in his account and the remaining Meditations are concerned with the processes of recovery (and, at the end of the book, with the potential threat posed by relapse), now that the true danger to Donne's life is passed. But the group of the bells Meditation (the sixteenth through the eighteenth) also serves as the culmination of Donne's spiritual—or, his *theoretical*—project that has sought through exemplarity to reimagine the problem of embodiment and insure the continuation of community against the threats posed by the isolation and fragmentation of the atomized world.[51]

III

As a final effort to characterize Donne's method as I have theorized it in this chapter, I would like to conclude with a very brief discussion of that alternative and contemporaneous early modern resolution to the question of method—a philosophical resolution that has both displaced Donne's version and helped to render it perhaps radically unfamiliar to "modern" readers. Though there are indeed significant correspondences and even similarities between these two models—both are orthodox Christian systems, both are exegetical in nature, both are involved in the invention of the scientific, and both are fundamentally autobiographical in articulation—Donne's meditational–theological–critical method and René Descartes's philosophical–scientific *Method* differ on one essential point: the problem of

---

shall in a blessed and glorious resurrection give me such an issue from this death as shall never pass into any other death, but establish me into a life that shall last as long as the Lord of Life himself. [p. 409]

[51] This notion that this set of Meditations culminates—and so the *Devotions* more generally—in this expression of the exemplary is a different conclusion from the suggestion found in some criticism that Donne's concerns here are merely the concerns of the egoist. Writing about the famous passage in Meditation 17 ("And therefore never send to know for whom the *bell* tolls; It tolls for *thee*" [*Devotions*, 87]), Targoff writes:

The gesture of the final "thee" is no doubt the rhetorical turn most responsible for this passage's enduring fame, and its power lies in its reach from the confines of Donne's sickbed to the innumerable "thee[s]" who will read his text. It is also a gesture of the egotistical sublime of which Donne is often accused: everything in his world can be refracted, absorbed into the universe of his own experience. [Targoff, 2008: 148]

thinking and its relation to the body. For while Donne's project in many ways is dedicated to thinking *through* the body, Descartes's, to the contrary, represents thinking *without* the body.

It would be no exaggeration to suggest that Descartes thereby establishes the dualism that Donne had sought so assiduously to pre-empt. Indeed, the *Devotions*—like Donne's work in poetry and prose more generally—is deeply invested in the ideal, which is to say the idealized, value of the human body. Although Donne's faith will not allow him to embrace any notion of corporeality other than the orthodox one that sees the flesh as inherently corrupt and corrupting, the fantasy that informs his work works in the opposite direction: toward an affirmation and legitimation of the body. Necessarily, of course, this legitimation can only happen (if at all) after the body's material dissolution, so graphically described in the *Devotions*, and only after God's active and deliberate restitution and purification of the body. The fourteenth Expostulation, to cite one important instance, gives voice to Donne's hopeful vision of his restored body when "this *day* of *death* shall deliver me over to my *fift day*, the day of my *Ressurection*":

> Then wee shall all bee invested, reapparelled in our owne *bodies*; but they who have made just use of their former *dayes*, be super-invested with *glorie*, wheras the others, condemned to their *olde clothes*, their *sinfull bodies*, shall have *Nothing* added, but *immortalitie* to *torment*. And this *day* of awaking me, and reinvesting my *Soule*, in my *body*, and my *body* in the body of *Christ*, shall present mee, *Bodie*, and *Soule*, to my *sixt day*, The day of *Judgement*. [*Devotions*, 76]

As powerful as this formulation is, however, and as consistent as Donne is in his reference to it or elaborations of it, it is finally not the method that ultimately resolves the matter of thinking and embodiment. That distinction belongs to Descartes, even if the philosophical dualism he champions resolves the question by way of a radical *dis*-embodiment.

Descartes begins his *Discourse on the Method* by locating the object of his study: setting aside oratory, poetry, mathematics, moral teachings, philosophy, theology, jurisprudence, medicine, and history, Descartes "entirely abandoned the study of letters" and resolves "to seek no other knowledge than that which could be found in myself or else in the great book of the world."[52] This grand act of rejecting all of the disciplines of knowledge represents a restriction of focus that is counterbalanced by an initial dilation: from books, let us say, to the entire world instead. But this movement outward is followed by its own countermovement, a constriction of that world: from the world writ large, to Germany, to a single village, to a single "stove-heated room, where I was completely free to converse with myself about my own thoughts" (p. 116). This constriction of Descartes's world to a single room allows for his discovery of the fundamentals of a method for attaining certainty in knowledge. Descartes presents these maxims, as he calls them in the *Discourse*, which are dedicated to the elimination of all "opinion" through the

---

[52] Descartes (1985: 115). Subsequent references are to this edition.

deployment of radical doubt. But this discovery leads in turn once again to a dilatory movement back into the world and, expecting "to be able to achieve this more readily by talking with other men than by staying shut up in the stove-heated room," Descartes embarks on nine years of travel, all the while "trying to be a spectator rather than an actor in all the comedies that are played out there" (p. 125). Traveling and watching, practicing the method and gathering experiences, and even earning the reputation of a man who had gained the sought-after wisdom, Descartes runs the course of these nine years, learning much (we are told), but deciding nothing "regarding the questions which are commonly debated among the learned" and taking no steps toward the "search for the foundations of any philosophy more certain than the commonly accepted one." This is work that requires another leave-taking and believing it necessary "to try by every means to become worthy of the reputation that was given me," Descartes retreats once again, this time decisively:

> Exactly eight years ago this desire made me resolve to move away from any place where I might have acquaintances and retire to this country [Holland], where the long duration of the war has led to the establishment of such order that the armies maintained here seem to serve only to make the enjoyment of the fruits of peace all the more secure. Living here, amidst this great mass of busy people who are more concerned with their own affairs than curious about those of others, I have been able to lead a life as solitary and withdrawn as if I were in the most remote desert, while lacking none of the comforts found in the most populous cities. [p. 126]

Unlike Donne, whose isolation is forced upon him by the ravages of disease and for whom solitude is a worse torment than hell itself represents, Descartes pursues his isolation deliberately, joyously, even. Donne is rescued from solitude by community and salvation comes in the form of an ever-expanding community: fellow Londoners, Christians, the dead, the saved, God. And all the while the body is fully present in both a material world and a material eternity.

In the *Discourse* (as in the later *Meditations on First Philosophy*, which covers a portion of this same ground), Descartes seems not to experience isolation as solitude, nor does society intervene into his isolation (except in the form of "comforts" the metropolis affords):

> But immediately I noticed that while I was trying thus to think everything false, it was necessary that I, who was thinking this, was something. And observing that this truth *"I am thinking, therefore I exist"* was so firm and sure that all the most extravagant suppositions of the sceptics were incapable of shaking it, I decided that I could accept it without scruple as the first principle of the philosophy I was seeking.
>
> Next I examined attentively what I was. I saw that while I could pretend that I had no body and that there was no world and no place for me to be in, I could not for all that pretend that I did not exist . . . From this I knew I was a substance whose whole essence or nature is simply to think, and which does not require any place, or depend on any material thing, in order to exist. Accordingly this "I"—that is, the soul by which I am what I am—is entirely distinct from the body, and indeed is easier to know than

the body, and would not fail to be whatever it is, even if the body did not exist. [p. 127]^[53]

Indeed, for Descartes, isolation is ever more rarified through a proliferating and ever-expanding leave-takings that serve to disembody the entire world.

Yet, at the end of the *Discourse*, Descartes has occasion to think about a particular embodiment. He tells his reader that he had, by means of his method, accumulated a great degree of insight into the physical world and the laws by which it is both constructed and maintained. "I endeavoured to explain the most important of these truths," he writes, "in a treatise which certain considerations prevent me from publishing" (pp. 131–2). In lieu of this unpublished treatise, Descartes can only offer, he says, a summary of its contents—and he proceeds to enumerate them: the nature of light and fire, cosmology, the function of the heart, the thinking soul, the eternal nature of the soul, animal spirits, the laws of mechanics, and a great many other discoveries. "It is now three years since I reached the end of the treatise that contains all these things," Descartes writes. "I was beginning to revise it in order to put it into the hands of a publisher," he continues, until news from Rome stopped him in his tracks. And the figure of Galileo standing before the Office of the Inquisition—and its instruments of torture—rises in Descartes's imagination:

> I was beginning to revise it in order to put it into the hands of a publisher, when I learned that some persons to whom I defer and who have hardly less authority over my actions than my own reason has over my thoughts, had disapproved of a physical theory published a little while before by someone else. [pp. 141–2]

And while Descartes says he had himself "noticed nothing in it [Galileo's 1632 *Dialogue Concerning the Two Chief World Systems*] that I could imagine to be prejudicial either to religion or to the state," he is sufficiently aware of the risks to body and soul that "That was enough to make me change my previous decision to publish my views" (p. 142).

The Copernican body, we can say, remains vulnerable to steel. And to ideology.

---

[53] Discussing this passage, as well as a related one from the *Meditations*, Johnson (2001) writes:

The cogito is grounded on the unshakable foundation of a thought without articulation, an unimpressionable thought that leaves no trace of itself, of the cogito it expresses and eposes without leaving behind. No body occupies the place where the cogito posits itself, which means there is no chance of the cogito ever taking another object for itself, of confusing and mistaking another subject for itself. There is no other there because the cogito is not there either. [p. 125]

6

Nature's Art

In this chapter I consider two responses to the debate of the art–nature relationship that reflect different epistemological trajectories in the early modern period. One response to the question of the ways in which art and nature are related is found in Shakespeare's late play *The Winter's Tale*. In this text, Shakespeare resolves the great contest through a characteristically imaginative maneuver: he will declare (in the words of Polixenes, and as ratified by Perdita) that "art itself is nature."[1] The second response is an antithetical one because it exactly inverts this formulation. In his magisterial *Elysium Britannicum*, John Evelyn will assert that the gardener is free to use artificial objects in the design and construction of the garden precisely because "Nature has already bin (as we may truly say) so Artificiall."[2]

I take Shakespeare's notion that art is nature to be typical of a certain conventional settlement of this ancient question. At the same time, and especially as it is developed in *The Winter's Tale*, this idea of the naturalness of art (if I can be granted the phrase) takes a particular form on the stage, the *trompe l'oeil*, which it will be the task of the first part of this chapter to discuss and analyze. Evelyn's equally striking, though diametrically opposed, assertion that nature is already artificial, for its part, takes on a particular form in his garden-writings. This form, as I will argue in the second part of this chapter, has among its aims the rejection of the discourse of the *trompe l'oeil* and, at the same time, the rejection of an entire emblematic epistemology that underwrites the Shakespearean settlement staged so powerfully in *The Winter's Tale*.

I

The Winter's Tale not only offers an elaborately staged theatrical *trompe l'oeil* in the famous statue vivification scene at the end of the play (if that is indeed what it is), but more importantly for this discussion, is *itself* an elaborate *trompe l'oeil* in which what is at stake is precisely the relationship of art to nature. *Trompe l'oeil* as a concept—or, perhaps, as a mechanism or machine—provides the structure of experience, whether visual or intellectual, for the contest between art and nature. And it does so by virtue of its own complicated structure of artifice in play with what is presumed to be nature—even if that presumed nature turns out in the end

---

[1] Shakespeare (1996: 4.4.97). Subsequent citations are to this edition of the play.
[2] Evelyn (2001: 58). Subsequent citations appear parenthetically.

to be artificial or even imaginary. In his seminar "Of the Gaze as *Objet Petit a*," Jacques Lacan speaks to this complex structure of the *trompe l'oeil* which functions only by virtue of seeming to promise something that it simultaneously refuses to manifest. But crucially, the trick (or, perhaps, the tease) only works once we understand it as a trick. "What is it that attracts and satisfies us in *trompe l'oeil*?" Lacan asks, not wholly rhetorically. "When is it that it captures our attention and delights us?" As Lacan will immediately clarify, the delight does not reside in the deception alone:

> At the moment when, by a mere shift of our gaze, we are able to realize that the representation does not move with the gaze and that it is merely a *trompe l'oeil*. For it appears at that moment as something other than it seemed, or rather it now seems to be that something else.[3]

The structure of the *trompe l'oeil* as Lacan theorizes it in the service of theorizing desire and its relation to presence/absence, is linked to Shakespeare's play in a number of ways.[4] Lacan's understanding of the *trompe l'oeil* can offer, for instance, a useful way of thinking about the vivification of the Hermione statue in Act 5. But I am interested here in a deeper connectedness. I have in mind the function of *trompe l'oeil* as a structuring device that provides the model by which *The Winter's Tale* appears to promise a number of things that it just as carefully refuses to manifest, thereby engendering a playfulness between the promise and its refusal that, after Lacan, we can say goes a long way toward describing how the play fascinates, satisfies, and delights.

The play offers a figure (in name only, however, and therefore only as an absence) that stands as an apt machine in which are consolidated a number of other promises that the play will not keep, other presences that the play will steadfastly maintain as absent. This crucial figure is the artist Giulio Romano. The play's reference to Giulio is famously Shakespeare's only reference to a "modern" artist. In the play, Giulio is said to have created the statue of Hermione that Paulina keeps in her chapel—"a piece many years in doing and now newly performed by that rare Italian master Giulio Romano, who, had he himself eternity and could put breath into his work, would beguile nature of her custom, so perfectly he is her ape" (5.2.93–8). The absolute lifelikeness of this statue is

---

[3] Lacan (1978: 112). Lacan will argue that the *trompe l'oeil* offers a glimpse of the *objet a* precisely in the gap between the two experiences of the *trompe l'oeil* (in this case) painting: between the instant of our deception and the instant of our realization of that deception; he continues:

> The picture does not compete with appearance, it competes with what Plato designates for us beyond appearance as being the Idea. It is because the picture is the appearance that says it is that which gives the appearance that Plato attacks painting, as if it were an activity competing with his own.
>
> This other thing is the *petit a*, around which there revolves a combat of which *trompe l'oeil* is the soul. [p. 112]

[4] Belsey (1995), in a compelling reading of Shakespeare's *Venus and Adonis*, argues that Shakespeare's narrative poem functions "as a kind of trompe l'oeil, moving undecidably between modes of address and sustaining the desire of the reader in the process" and that the text is "a poetic record of the originating moment of desire" (p. 258).

foretold—"He so near to Hermione hath done Hermione that they say one would speak to her and stand in hope of answer" (5.2.98–100)—even without the benefit of this witness having actually seen the thing itself.

The appearance of Giulio as a sculptor—rather than as an architect and painter, as he is known to us today—is curious and interesting and has prompted much debate. Stephen Orgel's clear-eyed response to this matter is certainly correct: pointing to Giulio's epitaph cited first by Giorgio Vasari that seems to identify Giulio as a sculptor—"*Videbat Jupiter corpora sculpta pictaque/Spirare aedes morta-lium aequarier coelo/Julii virtute Romani*' ('Jupiter saw sculpted and painted bodies breathe and the houses of mortals made equal to those in heaven through the skill of Giulio Romano')" [p. 221–2, n. 95]—Orgel concludes, "Therefore whether Giulio in fact was or was not a sculptor (and the epitaph is difficult to argue away), the problem as far as Shakespeare is concerned is surely solved: Giulio *was* known in the Renaissance as a sculptor" (p. 57).[5]

This still leaves the question, however, of the nature of Shakespeare's *use* of Giulio. For Orgel, Giulio's statue functions as a sign of what emerges as one of the play's important topoi: "evidence of things not seen" (p. 57).[6] As a theatrical device Giulio Romano functions I argue as an elaborate *trompe l'oeil* that in effect stages the absence of a prior story about art and the limits of representation. What is promised and nevertheless withheld in the Giulio *trompe l'oeil* is the ancient story of the competition between the painters Zeuxis and Parrahasius—the same story (let me note) upon which Lacan constructs his reading of the *trompe l'oeil* as a figuration of the *objet petit a*.

That Giulio is a substitute for the Zeuxis–Parrahasius story becomes clear when we refer to Shakespeare's source text for *The Winter's Tale*, Robert Greene's 1588 prose romance, *Pandosto*. Greene references the Zeuxis–Parrahasius story in a pastoral scene that from a certain perspective (and in what I would call a certain *negative* relation to Shakespeare's play) may come to appear obscene. A young prince, having confessed his love to a shepherdess—"'Why, then,' quoth he, 'thou canst not love Dorastus?'"—and having heard something of a puzzle in response—"'Yes,' said Fawnia, 'when Dorastus becomes a shepherd'"—changes his royal clothes for the "unseemly rags" of a shepherd.[7] He then reappears before the shepherdess:

> "If thou marvel, Fawnia, at my strange attire, thou wouldst more muse at my unaccustomed thoughts. The one disgraceth but my outward shape; the other

---

[5] Giorgio Vasari, *The Lives of the Artists* (qtd in Shakespeare [1996: p. 222, n. 25]).

[6] Orgel writes, "Giulio Romano's statue is, in the more literal sense, the evidence of things not seen, said to have been sculpted by an artist whose statues, if he did in fact make any, Shakespeare could have known only by reading or by hearsay, a work created out of pure inference from a narrative." This underscores the fact that even in the period in which "connoisseurship was rapidly developing and actively pursued," Shakespeare turned not to a gallery but rather to "the text that constitutes the beginning of art history" (Shakespeare [1996: 57]).

[7] Greene, *Pandosto. The Triumph of Time* appears in a modernized text prepared by Stanley Wells as an appendix to Shakespeare (1996: 234–74, 259). As Orgel notes, *Pandosto* was popular enough to go through five editions by the time of *The Winter's Tale*.

disturbeth my inward senses. I love Fawnia, and therefore what love liketh I cannot mislike. Fawnia, thou hast promised to love, and I hope thou wilt perform no less. I have fulfilled thy request, and now thou canst but grant my desire. Thou wert content to love Dorastus when he ceased to be a prince and granted to become a shepherd, and see, I have made the change, and therefore hope not to miss of my choice." [p. 260]

But Dorastus has misjudged and now he must hear unwelcome words, in the guise of conventional wisdom, from the shepherdess:

"Truth," quoth Fawnia, "but all that wear cowls are not monks; painted eagles are pictures, not eagles; Zeuxis' grapes were like grapes, yet shadows. Rich clothing make not princes, nor homely attire beggars. Shepherds are not called shepherds because they wear hooks and bags, but that they are born poor and live to keep sheep; so this attire hath not made Dorastus a shepherd, but to seem like a shepherd." [p. 261]

Fawnia's clever response is interesting for a number of reasons and although my ultimate concern with this exchange is the reference to Zeuxis, it is worth pausing over other significant features of Fawnia's reply—including, for example, the fact that unknown even to Fawnia herself, she indeed only *seems* to be a shepherdess and her true identity as a lost princess will be revealed before the end of the story. There is also the fact that Fawnia's reply is more formal than final: once Dorastus reads a lecture to her on the transitory nature of youth and beauty ("Those which disdain in youth," he warns, "are despised in age. Beauty's shadows are tricked up with Time's colours") and once he assures her of the honor of his intentions ("I love thee, Fawnia, not to misuse thee as a concubine but to use thee as my wife"), Fawnia "could no longer withstand the assault," we are told, "but yielded up the fort" and confesses her abiding love of Dorastus (p. 261). And there is the further fact that in pursuit of its argument that appearances can be deceptive, Fawnia's reply makes certain claims about the nature of representation—both painted images and actions that a man might play—that serve to strip representation of value and reveal representation merely as the play of appearances that inevitably leaves reality untouched and unaffected. For all of his love and theatrical costuming, Dorastus only manages *to seem* to be a shepherd.

But Fawnia's reference to Zeuxis suggests that perhaps there is more at stake in this critique of representation than the commonplace argument that appearances can deceive. Once again: "painted eagles are pictures, not eagles; Zeuxis' grapes were like grapes, yet shadows." Fawnia reminds us that there are no objects in paintings, only paint that is made to refer to objects that by definition can reside only outside of paintings. Paint skillfully applied can function as a kind of trope—a metaphor, let us say—that allows us to believe (in spite of what we know) that we see grapes in Zeuxis's painting, but painting nevertheless remains an entirely referential system.

Greene would have found the story of Zeuxis and his defeat in the famous painting contest with Parrahasius (a defeat, it is worth pointing out, that would seem to imply that Parrahasius was the greater painter, though it is Zeuxis whom Fawnia invokes) in Pliny's *Natural History*. As rendered in Philemon Holland's 1601 English translation, Parrahasius issues the challenge and Zeuxis responds with a marvel: "*Zeuxis* for proofe of his cunning, brought upon the scaffold a table, wherein were clusters of grapes so lively painted, that the very birds of the aire flew flocking thither for to bee pecking at the grapes."[8] Given the testimony of the birds, the testimony, that is, of nature, one might well expect Zeuxis to be assured of victory, but Parrahasius's response is of an altogether greater order:

> *Parasius* againe for his part to shew his workmanship, came with another picture, wherein he had painted a linnen sheet, so like unto a sheet indeed, that *Zeuxis* in a glorious bravery and pride of his heart, because the birds had approoved of his handy-worke, came unto *Parasius* with these words by way of a scorne and frumpe, Come on sir, away with your sheet at once, that we may see your goodly picture: But taking himselfe with the manner, and perceiving his owne error, hee was mightily abashed, & like an honest minded man yeelded the victory unto his adversary, saying withal, *Zeuxis* hath beguiled the poore birds, but *Parrhasius* hath deceived *Zeuxis*, a professed artisane. [p. 535]

Pliny's version of this story frames the critical question at hand—deception through art—by reference to Zeuxis and his understanding of how he has lost the painting competition: Parrahasius has deceived the artist and, unlike the artist who only managed to deceive the birds, has demonstrated thereby the greater ability. But, as Lacan has taught us, there is only one true *trompe l'oeil* painting in this story, and that is Parrahasius's painting of the curtain: the "example of Parrahsios makes it clear that if one wishes to deceive a man, what one presents to him is the painting of a veil, that is to say, something that incites him to ask what is behind it" (p. 112).[9] And Parrahasius succeeds so well because he has understood the lesson of the *trompe l'oeil*: that his painting of the curtain is predicated on Zeuxis's desire that such a curtain is in fact the promise of a presence waiting to be revealed.

In *The Winter's Tale*, Shakespeare's redaction of the classical painting competition is staged in Hermione's statue scene in act five, but he earlier provides something of a dramatic and philosophical prologue: the debate over the status of hybridized flowers and their complicated relationship to both art and to nature.[10] In the famous sheep-shearing scene of Act 4, Perdita, having taken on the "hostess-ship o'th'day" (4.4.72), greets her guests and distributes flowers to them, including to the disguised Polixenes and Camillo. She first hands them rosemary and rue,

---

[8] Pliny (1601: 535).

[9] Lacan dismisses any claim to *trompe l'oeil* Zeuxis might make:

If the birds rushed to the surface on which Zeuxis had deposited his dabs of colour, taking the picture for edible grapes, let us observe that the success of such an undertaking does not imply in the least that the grapes were admirably reproduced, like those we can see in the basked held by Caravaggio's *Bacchus* in the Uffizi. If the grapes had been painted in this way, it is not very likely that the birds would have been deceived, for why should birds see grapes portrayed with such extraordinary verisimilitude? There would have to be something more reduced, something closer to the sign, in something representing grapes for the birds. [pp. 111–12]

[10] Shakespeare also revises the dramatic setting found in *Pandosto* in which the entire issue of art and nature (as represented in the Zeuxis story) takes place in an exchange between the two lovers, Fawnia and the disguised Dorastus. In Shakespeare's version, the debate occurs between Perdita and the disguised Polixenes.

"these keep/Seeming and savour all the winter long" (4.4.74–5).[11] Polixenes believes he detects a note of a certain aptness of these "flowers of winter" (4.4.79) for men of his age. Perdita replies by noting the season—"Not yet on summer's death nor on the birth/Of trembling winter" (4.4.80–1)—and hence the unavailability of other flowers, except those she rejects categorically as "nature's bastards" and therefore unacceptable to her: "the fairest flowers o'th'season/Are our carnations and streaked gillyvors,/Which some call nature's bastards; of that kind/Our rustic garden's barren, and I care not/To get slips of them" (4.4.81–5). When asked why, her reply is aimed directly at the issue of art and nature: "For I have heard it said/There is an art which in their piedness shares/With great creating nature" (4.4.85–7). Polixenes, giving voice to an ancient argument, offers in response a lecture on the nature of art and on the nature of nature.[12] "Say there be," he says, granting the premise, but seeking to deny the conclusion:

> Yet nature is made better by no mean
> But nature makes that mean; so over that art
> Which you say adds to nature, is an art
> That nature makes. You see, sweet maid, we marry
> A gentler scion to the wildest stock,
> And make conceive a bark of baser kind
> By bud of nobler race. This is an art
> Which does mend nature—change it rather—but
> The art itself is nature. [4.4.89–97]

Perdita does not deny the logic of the argument ("So it is," is her instant reply [4.4.97]), but when instructed to "make your garden rich in gillyvors./And do not call them bastards" (4.4.98–9), she flatly refuses, and in doing so not only addresses definitively (at least for her) the question of the relationship between art and nature, but makes explicit her knowledge that this discourse of gillyvors is at the same time the discourse of human sexuality, marriage, and reproduction:

> I'll not put
> The dibble in the earth to set one slip of them;
> No more than, were I painted, I would wish
> This youth should say 'twere well, and only therefore
> Desire to breed by me. [4.4.99-103][13]

For Perdita, it would seem to be the better course to leave nature untouched by art, even if (as she grants) within the practices of grafting and cultivation "art itself is nature." With this gesture, the play seems to resolve the fundamental question informing the debate over the nature of art and its relationship to nature: even

---

[11] Compare this moment to Ophelia's complicated distribution of daisies, columbines, rosemary, rue, and fennel in Act 4 of *Hamlet*.

[12] Orgel notes the antiquity of this topos (in Shakespeare [1996: 46 and n. 172]) and refers readers to Frank Kermode's discussion in his introduction to the Arden edition of *The Tempest* (1954).

[13] Perdita concludes the exchange by issuing another assembly of flowers, this time mid-summer flowers—lavender, mints, savory, marjoram, and marigold—dedicated, she says, "to men of middle age" she takes her guests to be (4.4.108).

though they may disagree about the status or value of its productions, both Perdita and Polixenes understand that art (or artifice) is itself nothing more than a category of the natural. And while this may also seem to put an end to the possibility of thinking about any other response to the art–nature question, the gillyvors debate is not, of course, the play's final word on the relationship between art and nature.

We turn now to the matter of Giulio's statue of Hermione prepared to accept the idea that art is a species of the natural, the preparatory work for this having been effected in the gillyvors debate. Indeed, the final scene of the play provides the most spectacular of Shakespeare's musings on this ancient question of the relationship between art and nature. It also provides a local habitation for Shakespeare's revision of the Zeuxis and Parrahasius episode and the art–nature debate of which it is a part—a revision that Shakespeare achieves through the logic of the *trompe l'oeil*. If (as Orgel argues) Giulio's statue is indeed "the evidence of things not seen" (p. 57)—and as such bears a similarity to other "things not seen," such as the meeting of Leontes and Polixenes and the crucially important "recognition scene" between Leontes and Perdita toward which the play (like fate) has been moving— then a great deal comes to depend on the logic of the *trompe l'oeil*. In the final scene we are presented with a statue so absolute in its lifelikeness to the lost Hermione (even to the subtlety of her having aged in the sixteen years since we last saw her) that the assembled group—and Leontes in particular—stand astonished:

> O, thus she stood,
> Even with such life of majesty—warm life
> As now it coldly stands—when first I wooed her.
> I am ashamed. Does not the stone rebuke me
> For being more stone than it? O royal piece!
> There's magic in thy majesty, which has
> My evils conjured to remembrance, and
> From thy admiring daughter took the spirits,
> Standing like stone with thee. [5.3.34–42]

Gradually Leontes detects signs of life artfully rendered: "See, my lord,/Would you not deem it breathed, and that those veins/Did verily bear blood?" (5.3.63–5); "The fixure of her eye has motion in't,/As we are mocked with art" (5.3.67–8); and:

> Still methinks
> There is an air comes from her. What fine chisel
> Could ever yet cut breath? Let no man mock me,
> For I will kiss her. [5.3.77–80]

The (re)animation of Hermione ("'Tis time; descend; be stone no more; approach;/ Strike all that look upon with marvel—come,/I'll fill your grave up. Stir—nay, come away,/Bequeath to Death your numbness, for from him/Dear life redeems you" [5.3.99–103]) may well cause questions for viewers and readers, even as it does for Leontes who says that this restoration of Hermione "is to be questioned, for I saw her,/As I thought, dead, and have in vain said many/A prayer upon her grave" (5.3.139–41). But questions about the status of the statue (is it Hermione or a miracle or art; has Hermione been resurrected from the dead or has she been alive this whole time?) are unanswerable within the world of the play—in part because Pauline refuses to admit them ("There's time enough for that,/Lest they desire upon this push to trouble/Your joys with like relation" [5.3.128–30]), and in part because the play does not invite our speculation about exactly how Hermione may have passed her sixteen years while awaiting Leontes's adequate repentance. The former represents the dramatic push to suspend our inquiry, while the latter tells us something about the nature of Shakespeare's theater:

> Leontes is our guarantee that the two deaths are real: if Mamillius is dead, so is Hermione [since they were buried in the same tomb]; and by the same token, if Leontes is being deceived by Paulina about the reality of death, so are we being deceived by Shakespeare. What this means is not that at the play's conclusion, Hermione really is a statue come to life (we have the word of Hermione herself that this is not the case), but that Shakespearian drama does not create a consistent world. [Shakespeare (1996: 36)]

These questions are unanswerable precisely because according to the logic of art as nature, the questions do not apply. It is sufficient in Shakespeare's play that loss is redeemed—at least in part: Mamilius remains dead and sixteen spent years cannot be restored.[14] Such redemption that the play does allow is enabled by a new form of experience that renders art and nature indistinguishable. This is all we can know about Hermione as the play ends.

But if this statue (to follow Lacan's reading of *trompe l'oeil*) functions as a veil to incite the desire for what it only *seems* to conceal, and since the statue has made good on at least one promise and delivered Hermione, we may well wonder about what the statue as *trompe l'oeil* has promised but has *not* delivered. I would suggest that this other promise that remains undeliverable is integral to the structure of the *trompe l'oeil* and related in the model of a critique to the Shakespearean settlement of the art–nature debate. In fact, what the statue will not reveal works against the notion offered in *The Winter's Tale*—and staged to such spectacular dramatic effect in the statue scene—that art is nature. What lies behind the statue as *trompe l'oeil* is multiply framed: it is both the scene from *Pandosto* in which the two young lovers argue over the story of Parrahasius and Zeuxis, and it is also the scene that (as it were) lies behind Greene's: the episode in Pliny that offers us the story of the painters and their competition.

Pliny's *Natural History* is dedicated to the more or less comprehensive study of the collected wisdom concerning the natural world. As such, there are, as one would

---

[14] Redeemed in part: "Now here at the end of *The Winter's Tale* a dead five- or six-year-old boy remains unaccounted for." This is Cavell (2003) on the *partial* nature of redemption (his term is "accounting") in the play. He continues, noting the unredeemed nature of Perdita's lost sixteen years:

> Time may present itself as a good-humored old man, but what he speaks about in his appearance as Chorus in this play is his lapse, his being spent, as if behind our backs. Then is the moral that we all require forgiveness and that forgiveness is always a miracle, taking time but beyond time? Any of these things can be said, but how can we establish or deliver the weight or gravity of any such answer? [pp. 193–4]

expect, discussions ranging from geography to human physiology, from proto-zoological discussions of land and sea animals to catalogues of plants and their medicinal qualities, and from grapes and viniculture to metallurgy. And although there are indeed particular instances in which Pliny's focus encompasses what is more obviously the artifactual world, his discussion of Zeuxis and Parrahasius constitutes a remarkable extension of the domain of natural history—the domain, that is, of what falls under the category of the natural.

The immediate context for the painting contest episode is a more general consideration of painting. Pliny arrives at this topic, quite naturally (so to speak) by way of a long discussion of metals and minerals—the natural materials, that is, from which paint is manufactured. But there is a still more important context within which the Zeuxis episode appears: Pliny's essentially historical discussion of the eclipse of painting by sculpture. Pliny begins the thirty-fifth book by indicating that having surveyed metals and minerals, he will turn to another sort of consideration, "omitting nothing that is necessary or follows a law of Nature."[15] Pliny then addresses painting, "an art," he writes, "that was formerly illustrious, at the time when it was in high demand with kings and nations and when it ennobled others whom it deigned to transmit to posterity" (*Natural History*, 261). Pliny's tone is elegiac because of the collapse in the prestige of painting in his own moment, a collapse that was precipitated by a number of factors, including (he says) the rise in prestige of sculpture: "at the present time [painting] has been entirely ousted by marbles, and indeed finally also by gold" (*Natural History*, 261). This is a loss, Pliny implies, not only of a particularly exalted art form, but of history, as well:

> The painting of portraits, used to transmit through the ages extremely correct like-nesses of persons, has entirely gone out. Bronze shields are now set up as monuments with a design in silver, with only a faint difference between the figures; heads of statues are exchanged for others, about which before now actually sarcastic epigrams have been current: so universally is a display of material preferred to a recognizable likeness of one's own self. [*Natural History*, 263]

For Pliny, in the absence of art's fulfilling this obligation to history, statues become not only anonymous, but merely self-referential. They relate and refer to nothing other than themselves and therefore represent only themselves. What is therefore lost is not just the art form itself, but the *moral* purpose of that art. "Consequently nobody's likeness lives," he laments, "and they leave behind them portraits that represent their money, not themselves" (*Natural History*, 263). For Pliny, art has its greatest value when it represents actual individuals and preserves both their like-nesses and identity through time. Portrait painting achieved this end, but sculpture seduced us into believing that the value of the portrait resided literally in the material from which it was produced. Sculpture, in other words, tricked us into believing that the value of the art object lies wholly in the veil and that the veil simply *is* the art and therefore signifies or promises nothing more. This is as absurd

---

[15] For this passage (and the remainder of this chapter) I quote from Pliny (1968: 9: 261). Subsequent citations will appear parenthetically as *Natural History*, followed by page number.

a situation as would have emerged had Zeuxis fallen in love with Parrahasius's painted veil and never considered for a moment what might exist behind it. This is the absurdity of the failure to recognize the *trompe l'oeil*.

To misread *trompe l'oeil*, then, is to think that art *is* nature. In this regard, while Shakespeare's choice of Giulio Romano as the sculptor of Hermione's statue is reasonable and apt enough, it nevertheless at the same time works against itself— and against the notion that art is nature. For in addition to his fame as pupil to Raphael and his heir, Giulio was legendary for his skill in *trompe l'oeil* painting.[16] One spectacular instance of this skill is his *Sala dei Giganti* depicting Jove's destruction of the Giants, in Mantua's Palazzo del Te, an entire room both constructed and painted as itself a gigantic and elaborate *trompe l'oeil*.[17] Giulio's skill in *trompe l'oeil*, then, seems in the end to argue against the idea that art is nature. For the theoretical point is that *trompe l'oeil* is always a joke and therefore essentially unlike Shakespeare's quasi-theological mystification that is the Hermione statue scene at the end of the play. Giulio's *trompe l'oeil* paintings work because they always resolve themselves on the precise fundamental point that Shakespeare's statue of Hermione cannot admit: that art is not nature; art is art.

In contrast to the assertion that art is a species of nature that we find in *The Winter's Tale*, there emerges through the seventeenth century the antithetical notion: that nature is a form of art. In the following section of this chapter, I would like to consider one particular articulation of this notion: the process I will call the artifaction of nature as enacted in the garden-writings of John Evelyn. Perhaps the most appropriate place to begin is to consider the nature of the "natural" in the garden. This is an especially urgent matter—and an especially vexed one—in that the materials that serve to constitute the garden are, in some sense, the very objects of nature—whether animate (the tree, for example) or inanimate (a stone)—and therefore always simultaneously "natural" and, to the extent that they are *semiotically* deployed, artificial. It is the objective of the garden, after all, to construct for its viewer (or reader) the artificial experience of nature.

---

[16] Tigner (2006) notes that Giulio's "reputation was built upon his playful tricks" (p. 127). For Tigner, he is also important to Shakespeare's play as a designer of gardens and this, in turn, sets the stage for a discussion of "Hermione's transformation as a garden automaton" (p. 128). For another discussion of *The Winter's Tale* and automata, see also Maisano (2007).

[17] In the chapter on Giulio in his monumental *The Lives of the Artists*, Vasari (1991) describes this room in great detail—from its inception, through its construction and subsequent painting. Among the many details of the completed room, Vasari points to the site of the fireplace, strategically located and integrated into the composition of the painting:

In another part [of the painting/room] Giulio represented other giants upon whom are crashing down temples, columns, and other parts of buildings, creating among these arrogant creature great havoc and loss of life. And in this spot among the buildings crashing down was placed the fireplace for the room which, when a fire is lit, makes it seem as if the giants are burning, for Pluto is painted there fleeing towards the centre with his chariot driven by wizened horses and accompanied by the hellish Furies. [p. 372]

II

> [T]o define a Garden now, is to pronounce it *Inter Solatia humana purissimum*. A place of all terrestriall enjoyments the most resembling *Heaven*, and the best representation of our lost felicitie. It is the common Terme and the pit from whenc we were dug; We all came out of this parsly bed.
>
> John Evelyn, *Elysium Britannicum*

In his landmark essay "The structuralist activity," Roland Barthes revisits Hegel's historical fable on the persistence of human fascination with "the *Natural* in Nature" and a readily-perceptible (though as yet unnamed) presence in the natural world.[18] Hegel identifies the Greek response to this presence as the invention of the god Pan; Barthes renames this presence the frisson of meaning:

> According to Hegel, the ancient Greek was amazed by the natural in nature; he constantly listened to it, questioned the meaning of springs, forests, storms; without knowing what all these objects were telling him by name, he perceived in the vegetal or cosmic order a tremendous shudder [*frisson*] of meaning, to which he gave the name of a god: *Pan*. [p. 218].

For Barthes, the critical issue at stake in this quasi-historical discussion is less an anthropology of the human–nature relationship than it is the articulation of a counterpoint to his sustained definition of the critical and theoretical practice—he will call it an "activity"—of structuralism.[19] And while distinct from its intellectual forebears, structuralism nevertheless has a certain link with the desire for meaning experienced by the ancient Greeks as Pan and by Hegel as Spirit, even if those terms are notoriously imprecise, as indeed they are. For Barthes, the frisson of the "*Natural* in Nature" turns out not to be a meaning immanent in the world— neither god (or later, God), nor Spirit—but rather the act of the production of meaning itself:

> Subsequently, nature has changed, has become social: everything given to man is already human, down to the forest and the river which we cross when we travel. But confronted with this social nature, which is quite simply culture, structural man is no different from the ancient Greeks: he too listens for the natural in culture, and constantly perceives in it not so much stable, finite, "true" meanings as the shudder [*frisson*] of an enormous machine which is humanity tirelessly undertaking to create meaning, without which it would no longer be human. [pp. 218–19]

The early modern garden is an ideal locus for a consideration of the frisson that Barthes describes, for if the garden shudders, it does so because it is a machine and

---

[18] Barthes (1972: 214). Barthes refers to Hegel's discussion of the Greeks and their relation to Spirit; Hegel (1956: 234–5).

[19] On the nature of structuralism as an activity, Barthes writes, "the first thing to be said is that in relation to all its users, structuralism is essentially an *activity*, i.e., the controlled succession of a certain number of mental operations" (p. 214).

because the lines of social and cultural force that construct it are struck—or (more appropriately) they are *sounded*—and the resonance that results is the frisson of the "scientific" drive for meaning that the garden as artifice embodies and as "natural" object simultaneously obscures. The frisson of the garden stands as a powerful effect of the garden's hybridized nature as a *natural* artifact of human knowledge.

An "archaeology" of the garden (to invoke Foucault's vocabulary) would require consideration of a range of cultural practices. From among the many possibilities for such analysis (one could study, for instance, the politics of land use and land reform, or a given aesthetic theory), I will discuss here the emergence of the garden in relation to the waning of what can be called an *emblematical* epistemology and the emergence, in its place, of the scientific.[20] In the middle of the seventeenth century, and in the garden-working and garden-writing of men such as John Evelyn and others associated with the founding of the Royal Society, the notion of the garden as strictly an emblem of God's inscription of absolute theological meaning in the world is displaced—though neither immediately, nor completely—by an understanding of the garden as locus of the human production of meaning.

The early modern garden marks that moment in which the garden ceases to be strictly a matter of recreation or art.[21] At the same time, the broad set of cultural practices that in the very same historical moment are beginning to merge and coalesce into what we now identify as science serve not only to encourage the shift of the garden away from diversion and aesthetics, but help materially to construct the culture into which the garden will move. As I have argued throughout this book, what has typically been alleged as an exceptionalism for science—as that practice that best accesses natural fact and truth—is in fact the end-result of a complex historical process through which diverse practices become consolidated under the disciplinary name "science." It is only by way of the conceptualization of a unifying culture that these practices become detached from their organic locales and implicated within the growing ideology of science. Early modern science *itself* functioned as a machine for the consolidation of like practices and the privileged production of natural meaning. In other words, science *understood as a culture* is the first experimental product/production of science.

As even a cursory survey of garden history and garden practice in the early modern period will make clear, there was in fact no universal or abstract garden, but rather an astonishing array of particular kinds of gardens–kitchen gardens, "coronary" gardens, floral gardens, hanging gardens, botanical gardens, and medicinal (or "philosophical") gardens, to name a few of the most noted. For the purposes of this discussion, I offer the term "garden" in a theoretical sense and intend it to signify the deliberate and self-conscious construction of natural forms *for the sake of form* that in fact all of the particular kinds of gardens just mentioned have in common.

---

[20] For a powerful analysis of the politics of land use—and cultural representations of land use—see McRae (1996), especially his discussion of the discourse of "improvement" (Ch. 5).

[21] The literature on the subject of the relation between the garden and art is vast; see especially the following recent studies: Ross (1998), Miller (1993), and Hunt (1992); see also individual essays published in *The Journal of Garden History*. For studies of the early modern English garden, see Preston (2006) and Bushnell (2003).

And perhaps it is not too fanciful to suggest that in whatever particular form the early modern garden appeared, it always had as its ultimate referent (even if unstated), the original garden, the Garden of Eden: the "*Inter Solatia humana purissimum*" to which Evelyn refers in the passage quoted at the start of this section.

But before the garden can be a paradise, it is a place of crisis, a site of struggle, a battleground.[22] As Susan Stewart reminds us in "Garden agon" (her discussion of the controversial garden-work of the Scottish poet Ian Hamilton Finlay), "A garden is the wresting of form from nature."[23] The garden that serves as the primary focus of her analysis is Finlay's "Little Sparta," but her discussion invites our speculation about the *idea* of the garden more generally. Stewart's discussion provides a point of critical departure: from a consideration of a particular garden practice to a more general consideration of garden theory, from Finlay's "simultaneously intransigent and sublime" art to an examination of "the meaning of making and unmaking more generally" (p. 115).[24]

But if it is the place of crisis—the garden agon Stewart describes—the garden is also an activity *in* crisis over the issue of theory. Locating the making of gardens within the more familiar context of "landscape architecture"—the nineteenth-century term invented by John Claudius Loudon that today has come "to signal the unease and lack of focus with which the modern professional views its activities"—John Dixon Hunt begins his recent book, *Greater Perfections: The practice of garden theory*, with a discussion of the manifest disparity between an abundance of garden practice, on the one hand, and the more or less complete absence of garden theory, on the other.[25] Hunt argues that although the garden has a deep history as practice, as theory it has effectively no history at all. Hunt's sense of the current precarious state of garden practice/theory is primarily elegiac in nature, an understanding that arises from a sense of a lost past:

> Above all, it has largely lost touch with whatever conceptual or thoughtful understanding of its activities was available to earlier generations of practitioners, and it has lost touch, too, with gardens not as items to be designed and built but as models or ideas for larger enterprises. Landscape architecture is uncertain of its way and at the same time profoundly skeptical of intellectual demands upon it. [p. xi]

---

[22] In the Western European tradition, the *locus classicus* for negotiating what might be called the theological garden is, of course, the Garden of Eden. For an enlightening discussion of this, see Prest (1981).

[23] Stewart (1998: 111). The scope of this struggle is not limited to the practical difficulties inherent in (re)negotiating the objects in the natural world, but rather extends into a more thoroughly cultural and theoretical struggle:

> We are reminded of all the meanings accruing around the term *agon*—a place of games, lists, courses; a national assembly of contests; a struggle for life and death; a battle; an action at law or trial; the argument of a speech; agony or anguish of the mind; the contest of the rhapsode; the struggle to assert oneself. [p. 115]

[24] For other discussions of Finlay's garden, see Ross (1988), Ch. 7, and Bann (1981).

[25] Hunt (2000: 1). See also, Hunt (1997). Hunt has written or edited numerous studies of gardens and garden history; see also (1996), (1992a), (1992a), and (1976).

This skepticism, as Hunt characterizes it, is the contemporary gardener's inheritance bequeathed by a long history of practice at the expense of theory. "[T]here was never a body of specialists," Hunt laments, "to compose treatises specifically for what we have come to call landscape architecture, as Vitruvius did for architecture" (p. 3). This is so complete an historical neglect, Hunt argues, that it is reflected in the absence of any poetic models, as well:

> It is entirely typical of the traditional assumption that the topic of place-making [Hunt's term for landscape architecture] was the province or specialty of no particular specialist that the Latin poet Virgil, writing on matters of agriculture and husbandry in his *Georgics*, announced that, though the topic of gardening was within the scope of his topic, he would leave it to be taken up by others. [p. 3]

It was only after the Renaissance, Hunt suggests, that this "classical lacuna" became a "source of some concern," and only then because (for a complex of social and historical reasons that his book will address) this was a period in which gardens "began to assume great importance" (p. 3). Though there are many important primary texts in and through which one can trace this gradual emergence—Hunt cites, among others, René Rapin's 1665 Latin poem, *Hortorum Libri IV*, Charles Estienne and Jean Liebault's 1564 *La Maison Rustique*, and Jacques Boyceau's *Traité du jardinage selon les raisons de la nature et de l'art* (1638)—Hunt turns to the figure of John Evelyn, whose "'Elysium Britannicum' may perhaps be considered the first and only attempt to survey the whole territory of garden art, garden history, garden theory, and garden practice" (p. 4). For Hunt, Evelyn's garden-work constitutes simultaneously both the emergence *and* the disappearance of precisely the sort of theorizing of which the garden stands in such serious need today.

Hunt identifies the work of Capability Brown as that particular figuration of the garden that "stands for the moment when people began to lose any understanding of garden art in terms of representation. . . . A new insistence upon and taste for naturalness had effectively collapsed the distance between medium and message" (p. 81). By the end of the eighteenth century, Hunt argues, "the materials of [the garden's] medium were not sufficiently different from its objects," and concludes that "In seeming to yield up their traditional role of imitating nature, gardens became only what they represented—flowers, shrubs, trees, and so on" (p. 84). Hunt identifies this as the crisis—the "Brownian dilemma"—facing contemporary efforts (as few as they may be) to construct a meaningful garden theory: "the necessary medium of nature must not be allowed wholly to subdue the evidence that it is itself being imitated" (p. 114).

Although Hunt will ultimately reject structuralist critical work as insufficient to the demands of the kind of garden theory he hopes will emerge and fill the void history has left, I would argue that structuralism provides an important insight to the pressing issue Hunt highlights: the nature of imitation in garden practice. On Barthes's account, the goal of structuralist activity is "to reconstruct an 'object' in such a way as to manifest thereby the rules of functioning (the 'functions') of this object." As such, it follows that structure emerges in and through this critical activity as "a *simulacrum* of the object, but a directed, *interested* simulacrum, since the imitated object makes something appear which remained invisible or, if one prefers, unintelligible in the nature object" (p. 214). In other words, structuralism stands as:

> essentially *an activity of imitation*, which is why there is, strictly speaking, no *technical* difference between structuralism as an intellectual activity, on the one hand, and literature, in particular, art in general, on the other: both derive from a *mimesis*, based not on the analogy of substances (as in so-called realist art), but on the analogy of functions (what Lévi-Strauss calls *homology*). [p. 215]

On this reading, the garden stands not simply as a material or "natural" analogy of the world, but rather as its homological counterpart. The conventional historicist understanding of the garden, in other words, never offers a way of thinking beyond the analogy. In this regard, such a mimetic theory of the garden shares what I will call an "emblematic epistemology" with the history of garden history itself. This is all the more striking in that it is precisely the conception of the garden as an emblem—and an informing and enabling emblematic epistemology—that is monumentally revised in the mid-seventeenth century by Evelyn and other Royal Society figures. This revision, as I argue below, is simultaneously underway in literary culture in the period. Even as natural philosophy begins to move away from the emblematic epistemology it had inherited, so, too, does the literary emblem tradition begin a shift from the pastoral tradition toward a new georgic practice. Though of course neither of these moves is either instant or complete; but, taken together, they help contribute to the emergence of the new idiom of science.

III

In his article "'Wild pastorall encounter': John Evelyn, John Beale and the renegotiation of pastoral in the mid-seventeenth century," Douglas Chambers traces what he identifies as the "shattering of the arcadian ideal of Stuart politics [that] inevitably brought with it the destruction of its pastoral mythology and demanded a renegotiation of pastoral with georgic."[26] Although I will want to describe more particularly mid-century georgic as the site for the inscription of the discourse of science, I would like here to mark Chambers's argument that through the middle years of the seventeenth century, men such as Beale and Evelyn (as well as others associated with the Royal Society) together forged a new version of the georgic: "the translation of the Virgilian text into the English landscape" (p. 181). This effort would lead from the founding of the Royal Society's Georgical Committee in the 1650s and 1660s to the "georgic arcadia" Chambers describes and that would be offered by eighteenth-century figures such as Alexander Pope (p. 181).[27]

---

[26] Chambers (1992: 176).
[27] Chambers continues:

> The landscape that Pope affirms in the "Epistle to Durlington" is one where use "sanctifies expense" and where "laughing Ceres" repossesses the land from the sumptuous excesses of Timon's villa. Here, as with Beale and Evelyn, it is ethics not economics that govern agriculture and horticulture: principles firmly grounded in the *Georgics*. [p. 187]

This same impetus toward a renegotiation of the pastoral that Chambers discusses is at the same time working its transformative effects upon another literary—and, I would add, epistemological—genre: the emblem. In response to the non-realism of pastoral (for which Chambers has described its political dimension and to which I will add its theoretical dimension), poets turn to a renewed form: the georgic. The question of the nature of the relationship between the pastoral and the georgic has been a much-debated one[28] and while my interest here is not in contributing one argument or another to it (indeed, I will hardly enter the debate), I would like to consider for a moment an argument offered by Alastair Fowler on the eclipse of the pastoral by the georgic through the seventeenth century. Fowler offers a corrective to critical discussions that tend to construe the relation between given genres as necessarily binary, suggesting that the history of the pastoral and the georgic offers one important instance in which literary forms—especially what he calls "neighboring genres"—blur together and their distinctness becomes ever more indistinct.[29] But rather than assigning responsibility to immutable and perhaps largely unknowable "laws" internal to genres themselves ("Changes in the generic hierarchy do not take place in a vacuum"), Fowler argues that it is necessary to situate the relative prominence or decline of genres within "the historical setting and shifts in domains of assumption, both literary and extraliterary" (p. 84). Accordingly, Fowler charts the emergence of the georgic as a major form against the historical moment of its changing fortunes from the sixteenth to the seventeenth century, including, importantly, the emergence of georgic from its Elizabethan incarnation as an ideal form for allegory. Similarly, even as in literary forms the impetus to medievalize began to wane, so too in the "extraliterary" world the political and economic organization of culture shifted away from the feudal or neo-feudal model: "with the larger norm for estates," Fowler suggests, "georgic mimed a more important national role: and the chivalric hero was replaced by a new civil ideal, the 'gentleman'" (p. 87).[30]

---

[28] The protestations of Marvell's "Mower" poems, for example—especially "The Mower against Gardens" in Marvel (1990: 40–1)—animate the line of confrontation between the pastoral and the georgic: the pastoral as the truly natural world, while the garden is construed as the site where nature is adulterated through the insidious interventions of the gardener.

[29] Fowler (1992); Fowler writes:

[I]t is precisely through fusion and confusion of genres that the canonic system achieves each new fixity. Thus, before georgic could ascend the hierarchy, it had to undergo a phase of mixture with pastoral. As we have seen, the impurity of English pastoral, gradually established over centuries, facilitated this temporary mixture: georgic was able to enter mixtures, change places, and within the space of less than a century achieve its exalted Augustan status. [p. 87]

[30] McRae's (1996) reading of what he calls "georgic economics" takes the development of the literary georgic (and Fowler's reading of this transformation in genre) as a point of departure; he will, however, move in an altogether different direction:

My main disagreement with the existing studies, however, centres on their tendency to detach literary production from economic practices and discourses ... I will adopt here a broader conception of culture and cultural influence, and argue that in the course of the period English writers produced a discourse which endorses the energy and diligence of the agrarian improver. By the middle of the seventeenth century 'georgic economics' consistently bound the expansive aims of the individual to a celebratory vision of national development. [pp. 199–200]

If Fowler's assertion regarding the production of the figure of the "gentleman" represents part of the social transformation of the seventeenth century, then Steven Shapin's *A Social History of Truth*, a study of the place of "civility" in the production of early modern science, constitutes a major articulation of precisely this move toward what we can tentatively call "modernity."[31] For Shapin, the emergence of early modern science is inextricably linked with this figure of the gentleman as the very embodiment of civility—that quality, based upon birth, wealth, and the accompanying notion of disinterestedness, that serves to enable the production of socially determined notions of both trust and truth. In the seventeenth century, "[t]he condition of securing knowledge about the nature of nature was the possession of knowledge about the nature of people" (p. 258). In a world in which "in certain sorts of people credibility was *embodied*," notions of credibility, reliability, and truth came to reside in the figure of the gentleman—that figure who, by virtue of social and economic station, functions in the world as an entirely "free actor," whose actions, including (importantly) testimony, are never constrained by material wants and who therefore acts out of more or less absolute disinterestedness (p. 238).

The special focus of Shapin's study is the figure of Robert Boyle who exemplifies—both in his own period and in Shapin's analysis—the ideal figure of the experimental philosopher: a gentleman who, in fact, goes a long way in establishing the seventeenth-century notion of the Christian virtuoso, and who works selflessly on behalf of humankind in the general project of securing scientific knowledge that is by definition *moral* in nature. In Boyle, Shapin argues, we can detect the salient features of early modern scientific practice: disinterested, selfless, objective, humble, practical, and moral knowledge about the material (and mechanical) world. To the extent that the gentlemanly experience of truth—or, the gentlemanly production of truth—served as the seventeenth-century "constitution of scientific truth," Shapin concludes that "Insofar as experience is obliged to transit a nexus of trust in order to become a part of knowledge, then there is no aspect of our knowledge we can speak about which can be set apart from our moral order" (p. 21). Shapin's notion of the "moral" nature of the production of scientific knowledge seems to me liable to the very sort of theoretical mistake Shapin condemns in more traditional studies of early modern science in that it tends to reproduce the "moral" as a transcendent category, rather than as a fully historical or contingent notion that depends upon emergent sociality. One could say that Shapin's treatment of civility and early modern science, while it wants to argue that the notion of "truth" has, as he says, a "social history," seems to deny a corresponding social history of the moral. It is this residual transcendent idealism, I want to suggest, that renders Shapin's argument problematical on the level of theory.[32]

At the same time that it attaches to the figure of the gentleman, the seventeenth-century rise of the georgic corresponds to what Fowler calls the "information explosion of the seventeenth century . . . [as] a large circumstance favourable to

[31] Shapin (1994).
[32] For a critique of Shapin's argument on the level of history, see Shapiro (2000, esp. 119–24).

the miscellaneous variety of georgic and to its appetite for factual details" (p. 84). Taken together, I would suggest, these two factors—the newly articulated figure of the gentleman as civil ideal and the "information explosion" and its effects—are instrumental in the production of early modern scientific culture.

There is a further important factor that contributes to the formation of scientific culture: a certain privileging of nonfictionality that emerges from within the discourse of the emblem. In his brief account of the ascendency of the georgic in the seventeenth century, Fowler takes Anthony Low to task for misreading Sidney's disregard for the georgic as simply a matter of a reputed aristocratic aversion to labor; Sidney, Fowler counters, "gave georgic a low place for quite another reason: because it lacked fictionality" (p. 86). It is this "nonfictionality" that will come to characterize both the georgic and the scientific idioms, at the heart of which lies a critique of representationalism inherent in the emblem itself. We can see such changes at work in the emblem poems of the controversial figure of George Wither.[33] In 1635, Wither (1588–1667) published a book of poems entitled *A Collection of Emblemes, Ancient and Modern*.[34] In his dedication to "the MAJES-TIE of Great *Britaine*, *France*, and *Ireland*, the Most Illustrious King, CHARLES; And his excellently beloved, the most gratious *Queene* Mary," Wither makes clear his understanding of the nature of the emblem as a vehicle that communicates a moral *visually*: "these/Are EMBLEMES, whose intention is to please/And profit vulgar Iudgement (by the view,/Of what they ought to follow, or eschew)" (*3). Among the two hundred didactic poems in the collection Wither has many that are dedicated to georgic labors. In the poem that accompanies the emblem depicting planting—"When I behold the Havocke and the Spoyle" (Figure 6.1)—Wither offers a version/vision of the life of the farmer and, at the same time, his praise of the moral virtues to be ascribed to the laborer who toils in the virtuous work of agricultural and animal husbandry:

> But, let these carelesse *Wasters* learne to know,
> That, as *Vaine-Spoyle* is open *Injury*;
> So, *Planting* is a *Debt*, they truely owe,
> And ought to pay to their *Posterity*.
> *Selfe-love*, for none, but for it selfe, doth care;
> And, onely, for the present, taketh paine:
> But, *Charity* for others doth prepare;
> And, joyes in that, which *Future-Time* shall gaine.

---

[33] For a discussion of Wither's career—including his alleged Puritanical sympathies and identification, his strife with the Stationers' Company, and Jonson's parody of him in the figure of Chronomastix in *Time Vindicated* (1623)—see the DNB entry; on the latter, see also Orgel (1969: 502–4).

[34] *A COLLECTION OF EMBLEMES/Ancient and Moderne/Quickened/With Metricall IL-LVSTRATION, both/Morall and Divine: And disposed into/LOTTERIES,/That Instruction, and Good Counsell, may be furthered/by an Honest and Pleasant Recreation./By GEORGE WITHER/The First Booke/LONDON,/Printed by A.M. for Robert Milburne, and/are to be sold at the Grey-hound in Pauls Church-/yard. MDCXXXV.* For a detailed discussion of the printing history of Wither's collection, including a discussion of the engraver and the first printing of the images for Gabriel Rollenhagen's *Nucleus emblematum selectissimorum* (Utrecht, 1611 [?] and 1613), see Freeman (1975: vii–xix).

He that delights to Plant and Set,
Makes After-Ages in his Debt.

35

ILLVSTR. XXXV.   Book. I.

When I behold the Havocke and the Spoyle,
Which (ev'n within the compasse of my Dayes)
Is made through every quarter of this Ile,
In *Woods* and *Groves* (which were this Kingdomes praise)
And, when I minde with how much greedinesse,
We seeke the present Gaine, in every thing;
Not caring (so our *Lust* we may possesse)
What Dammage to *Posterity* we bring:
They doe, me-thinkes, as if they did foresee,
That, some of those, whom they have cause to hate,
Should come in *Future-times*, their Heires to be:
Or else, why should they such things perpetrate?
For, if they thinke their *Children* shall succeed;
Or, can believe, that they begot their *Heires*;
They could not, surely, doe so foule a Deed,
As to deface the *Land*, that should be theirs.
What our *Forefathers* planted, we destroy:
Nay, all Mens labours, living heretofore,
And all our owne, we lavishly imploy
To serve our present *Lusts*; and, for no more.
    But, let these carelesse *Wasters* learne to know,
That, as *Vaine-Spoyle* is open *Injury*;
So, *Planting* is a *Debt*, they truely owe,
And ought to pay to their *Posterity*.
Selfe-love, for none, but for it selfe, doth care;
And, onely, for the present, taketh paine:
But, *Charity* for others doth prepare;
And, joyes in that, which *Future-Time* shall gaine
   If, *After-Ages* may my *Labours* blesse;
   I care not, *much*, how *Litle* I possesse.
F 2   To

Figure 6.1 "When I behold the Havocke and the Spoyle", George Wither, *A Collection of Emblemes, Ancient and Moderne* (1635)

If, *After-Ages* may my *Labours* blesse;
I care not, *much*, how *Litle* I possesse.
[Book 1: 35, ll. 21–30]

Wither's emphasis on the farmer and the efforts required to win a living from the
land represent a certain investment in something like poetic realism: unlike the
more clearly fictional discourse of the pastoral—a genre dedicated to a wholly
rarified natural world unnaturally devoid of work—these poems strive to make clear
that the virtuous life it means to depict depends upon work and effort. In this
regard, Wither's georgic poems stand in clear opposition to the highly stylized
notion of the natural world articulated, for example, in the great country-house
poems of the earlier decades of the seventeenth century—a poem such as Jonson's
"To Penshurst," with its version of the natural world that spontaneously offers its
vast riches for human consumption (to say nothing of the idealized and rigorously
fictional version of pastoral one encounters in *The Winter's Tale*). By contrast,
Wither's poems describe a world predicated upon acts of human labor—labor that
is in the first instance required to win from nature the materials necessary to survival
and, in the second, labor that serves as a model for the work of the soul in its
journey to God. Wither's text and his salvational ideology come to depend upon
this labor and it is in part upon the idea of labor that the (new) georgic stakes its
claim to cultural and political relevance.

In "*The Spade*, for *Labour* stands," Wither makes explicit this relation between
earthly labor and eternal salvation:

The *Spade*, for *Labour* stands. The *Ball with wings*,
Intendeth *flitting-rowling-worldly things*.
This *Altar-stone*, may serve in setting foorth,
Things firmer, sollid, and of greater worth:
In which, and by the *words* enclosing these,
You, there may read, your *Fortune*, if you please.
If you, your *labour*, on those things bestow,
Which *rowle* and *flutter* alwayies to and fro;
It cannot be, but, that which you obtaine,
Must prove a *wavering*, and unconstant gaine:
For, he that soweth *Vanitie*, shall finde,
At *reaping-time*, no better fruit than *Winde*.
. . .
Of *Sexe*, or of *Degree*, there's no regard:
But, as the *Labour*, such is the *reward*.
To *worke-aright*, oh, *Lord*, instruct thou me;
And, ground my *Workes*, and *buildings* all on thee:
That, by the fiery *Test*, when they are tride,
My *Worke* may stand, and I may *safe* abide.
[4.239, ll. 1–12, 25–30][35]

[35] Low (1985) argues that it was precisely this idea of labor and its idealization that prompted
members of the English aristocracy, such as Sir Philip Sidney, to repudiate the georgic in favor of the
apparently more conservative pastoral against which to a large extent the georgic had—and perhaps
has—historically been defined (pp. 3–34).

Figure 6.2 Author's portrait, George Wither, *A Collection of Emblemes, Ancient and Moderne* (1635)

Labor stands as the mark of realism—or nonfictionality—and upon this claim to realism the georgic poet leverages the repudiation of fictionality that underwrites the pastoral. And yet, there is a critical tension evident in the emblematic philosophy of Wither's text between his georgic realism, on the one hand, and a skeptical view of representation, on the other. We see this agon (to borrow Stewart's term) in Wither's complex set of opening gestures. Wither prefaces his *Collection* with an engraving of himself (with the Latin phrase "GEORGII WITHERI POETÆ

EFFIGIES" inscribed in the oval frame of the image), and the verse motto subscribed, "*What I WAS, is passed by;/What I AM, away doth flie;/What I SHAL BEE, none do see;/Yet, in that, my BEAUTIES bee*" (*3v) (Figure 6.2). This is followed by a poem, "The AVTHORS Meditation upon sight of his PICTVRE" in which Wither comments first upon what is finally the fictive and insubstantial nature of pictures:

> A PICTVRE, though with most exactnesse made,
> Is nothing, but the SHADOW of a SHADE.
> For, ev'n our living Bodies, (though they seeme
> To others more, or more in our esteeme)
> Are but the shadowes of that Reall-being,
> Which doth extend beyond the Fleshly-seeing;
> And, cannot be discerned, till we rise
> Immortall-Objects, for Immortall-eyes. [A4][36]

The logic of Wither's Christian emblematical conception requires an understanding of the transitory nature of worldly appearance and, indeed, objects themselves, pending their translation (perfection) in paradise. At the same time, the articulation of even this formulation (to say nothing of the entire collection of poems it introduces) completely depends upon a faith in the referentiality of image in order to work at all. What Wither's text demonstrates is that, to the extent that it depends upon a fundamental nonfictionality of the image, the emblem nevertheless carries within itself the controversy of representation—particularly when that emblem is indexed to a Christian theology.

The effects of this tension within the emblematical tradition—both poetic and epistemological—have substantive consequences. Among these is the emergence of a renewed interest in the georgic, not only as a poetic form but perhaps more importantly as an epistemology. Unlike the pastoral, the georgic idiom is wholly dedicated to the articulation of a material realism, albeit one articulated in verse or in science, neither of which is in any way natural or any less constructed, nor any more in an immediate relation to the natural world. The appearance, one could say, of such (fictive) immediacy in both verse and in scientific discourse is one of the collateral effects of the georgic idiom that posits itself heroically as nearer the natural world, though this nearness is nothing more, in the end, than a constructed proximity, nothing more than an effect of a rhetorical or a discursive set of

---

[36] Wither continues,

> For, as I view, those Townes, and Fields, that be
> In Landskip drawne; Even so, me thinks, I see
> A Glimpes, farre off, (though FAITH'S Prospective glasse)
> Of that, which after DEATH, will come to passe;
> And, likewise, gained have, such meanes of seeing,
> Some things, which were, before my Life had being,
> That, in my Soule, I should be discontent,
> If, this my Body were, more permanent;
> Since, Wee, and all God's other Creatures, here,
> Are but the Pictures, of what shall appeare. [A4v]

operations. Indeed, early modern science is *itself* an embodiment of the georgic idiom and through the seventeenth century we can see the gradual emergence of an understanding (philosophy) that can be said to run, in effect, parallel to the long-standing desire and practice of reading the material world emblematically. Rather than seeing nature as strictly an emblem of God and his Providence (in the model afforded by natural theology, which not only pre-dated the early modern period but in fact persisted long after, reaching something of a culmination in post-Reformation culture), early modern science enabled an understanding of the material world full of "things" rather than "signs."[37]

In his book *The Word of God and the Languages of Man: Interpreting nature in early modern science and medicine*, James Bono takes up precisely this issue of the epistemological reconstruction of the material world in the early modern period. Bono's rich analysis is articulated along the axes of (1) the gradual eclipse of hermeticism and the occult by early modern science, and (2) competing theories of language that can be said to underwrite both hermeticism and science. One can detect in the works of such seventeenth-century figures such as Bacon, Harvey, and Descartes, Bono writes, a reconceptualization of attempts to read the "book of nature": rather than relying upon the humanist practice of exegetical hermenuetics, in which the natural world is understood as a complex semiotic system designed and written by God and in which the image of God is embedded in hidden or occult fashion, early modern science moves instead toward a new interpretive system in which the world is no longer imagined as full of God's signs in the guise of the objects of nature, but rather is understood as full of "things" themselves—a transformation Bono characterizes as the shift from "abstract symbolism" (identifiably at work for figures such as Pico and Ficino) through "symbolic literalism" (seen in such figures as Paracelsus and Fernal), to, finally, the practices of reading nature as an order of contingent things, such as in the works of Bacon and Harvey.[38] Where exegetical hermeneutics tended analogically toward an underlining and ultimately discernable and knowable grand unity of creation, Baconian contingent reading of "things" works toward a greater understanding—metonymic in nature—of the great diversity of creation. This process, which comes to characterize the various practices of early modern science, constitutes a complex system of "de-inscription": the systematic (and eventually the institutionalized) attempt to detect God's inscription of order in the objects of the world (pp. 83–4). In this system, there is simply no place for the occult and hermetic theories of a Paracelsus. Instead, we can see in the works of Harvey, for instance, the radical rejection of such (humanist) notions as transcendent causality: where Fernal wrote extensively about the "medical spirits" he believed responsible for animation, Harvey will argue an immanentist interpretation that locates "spirit" within the very elemental object under consideration—his theory that it is blood itself that animates life, free from any super-elemental (or divine) addition:

---

[37] For a helpful discussion of natural theology (and its relation to the practices of science), see Peterfreund (2008) and (2000).
[38] Bono (1995: 81).

There are I say mixt and compounded bodies, even in respect of time before any Elements, as they call them, into which they are corrupted and determine; for they are dissolved into those Elements rather in order to our apprehension, then really and actually. And therefore those bodies called Elements, are not before those things which are made and generated; but rather after them, and their Reliques rather then their Principles.[39]

The work of Harvey can stand here as generally representative of the profound shift in the epistemology (to say nothing of the theories of language so important to Bono's study), an epistemology that characterized seventeenth-century science and its move toward a conceptualization of what Bono calls "the nonsymbolic text of nature" (p. 193). It is only once this innovation of "de-inscription"—at once both philosophical (in theory) and methodological (in practice)—is secure that the culture can move toward a scientific understanding of nature:

> The metonymy of identity and difference, of contiguity and displacement—ultimately of cause and effect—thus replaces for Bacon all traces of the metaphorics of resemblance, correspondence, and emblematic meaning. The discourses of types . . . gives way to a discourse of order and diversity, in which the very multiplicity of things becomes the subject of scientific inquiry, and the ordered table of natural history becomes the objectified artifact of human knowledge and of the divine Book of Nature. [p. 244]

The garden-works of John Evelyn constitute one locus for this process of the "artifaction of nature"—that process by which the natural world *itself* is available to scientific inquiry only once it has been submitted to and transformed by its insertion into an epistemological process for the production of meaning.[40] Evelyn's horticultural ethics of nonfictionality is clearly dedicated to the careful and (eventually) the complete elaboration of a practical or experimental philosophy of nature in the georgic idiom.

IV

Though best known to us by his *Kalendarium* as the great diarist he indeed was, in his own long and productive lifetime (he lived to the age of 86 and wrote daily to within two or three weeks of his death in 1706), John Evelyn's widespread fame was based largely upon his work on the garden. Evelyn's interests were both practical and theoretical in nature, leading him, on the one hand, to the construction of his own famous and important gardens at Sayes Court. On the other hand, this interest in and devotion to the garden as an *intellectual* enterprise led Evelyn to write a series of landmark texts, including *Acetaria: A Discourse of Sallets*;

---

[39] William Harvey, *Anatomical Exercitations Concerning the Generation of Living Creatures* (1653), 468; qtd. in Bono (1995: 122).

[40] For a discussion of the processes of artifaction in the early modern period more generally, see Marchitello (1997).

Kalendarium Hortense; *Terra: A Philosophical Discourse of Earth*; *Sylva*; and, his hortulan *magnum opus*, the incomplete and (until the year 2000) unpublished *Elysium Britannicum*.[41] Evelyn characterized his *Elysium Britannicum* (a work that has been called "not simply the most important unpublished document in English garden history, [but] one of the central documents of late European humanism"[42]) as the work of "almost 40 years" spent:

> in gathering and amassing Materials for an Hortulan Design, to so enormous an Heap, as to fill some Thousand Pages, and yet be comprehended within two or three acres of Ground; nay within the Square of less than One (skillfully planted and cultivated) sufficient to furnish, and entertain [my] Time and Thoughts all [my] life long, with a most Innocent, Agreeable, and Useful Employment.[43]

Evelyn's garden activities can be seen as manifestations of the ambition—one shared by many leading intellectual figures of the period—to further the cause of experimental philosophy. Along with such illustrious friends as John Beale, Samuel Hartlib, and Robert Boyle, Evelyn was one of the founding members of the Royal Society, an institution dedicated from its very inception to what Evelyn calls "Real Philosophy." In the prefatory letter to *Acetaria* dedicating the book to John Somers, Lord Chancellor of England and President of the Royal Society, Evelyn speaks to this passionate devotion to practical and useful knowledge:

> Thus, whilst King *Solomon*'s Temple was *Consecrated* to the *God* of *Nature*, and his true Worship, This may be Dedicated, and set apart for the Works of Nature; deliver'd from those Illusions and Impostors, that are still endeavouring to cloud and depress the true and *substantial Philosophy*: A *shallow* and *Superficial Insight*, wherein (as that Incompa-rable Person rightly observes) having made so many *Atheists*: whilst a *profound* and thorow *Penetration* into her *Recesses* (which is the *Business* of the *Royal Society*) would lead Men to the *Knowledge* and *Admiration* of the *glorious Author*. [*Acetaria*, 135]

In his prefatory letter "To the Reader," at the opening of the fourth edition of *Sylva* (1706), Evelyn attempts to make clear the fundamental distinction between an older model of inquiry, integral to scholasticism, on the one hand, and his own pragmatics, aligned with the efforts of the Royal Society, on the other: it is the work of the Royal Society, Evelyn declares, to produce what he called "*Real Philosophy*." In order to fulfill its mission "to *improve Natural Knowledge*, and inlarge the *Empire* of *Operative* Philosophy; not by an Abolition of the *Old*, but by the *Real* Effects of the *Experimental; Collecting, Examining*, and *Improving* the scatter of *Phenomena*," the active and practical—the pragmatic—natural philosophy undertaken by the Royal Society must act, Evelyn suggests, as the good architect would: with the "pulling down [of] the decay'd and stinking wall to erect a better, and more substantial in its place" (*Sylva*, **2). In the process, the new-model natural

---

[41] For a detailed bibliographical survey of Evelyn's works, see Keynes (1968). There is also a recent edition of selected works (including *Sylva*, 1664 and *Kalendarium Hortense*, 1706): de la Bédoyère (1995).

[42] Chambers (1997: 107).

[43] *Acetaria*, published in *Sylva*, 4th edn (1706: 139).

philosopher had to undertake the arduous labor of sifting through the products of scholastic inquiry:

> to see if they could find any thing . . . *sincere* and *useful* among this *Pedantick Rubbish*, but all in vain; here was nothing *material*, nothing of moment *Mathematical*, or *Mechanical*, and which had not been miserably *sophisticated*, on which to lay the stress; nothing in a manner whereby any farther *Progress* could be made, for the *raising* and *ennobling* the *Dignity* of *Mankind* in the *Sublimist Operations* of the *Rational Faculty*, by clearing the *obscurities*, and *healing* the *Defects* of most of the *Phisiological Hypotheses*, repugnant, as they hitherto seemed to be, to the *Principles* of real *Knowledge* and *Experience*. [*Sylva*, **2]

Evelyn's strategy, as these passages describe it, requires that such natural philosophers as gravitate to the Royal Society must take for their object of study the very works of nature themselves, that the study of the objects of nature itself constitutes (albeit in another form) labor that is duly consecrated to God and an expression of their worship of him, even if in a mediated fashion. This suggestion constitutes an important articulation of the shift away from reading the world strictly as an emblem of God's omnipotence. For Solomon, Evelyn says, the temple (which stands for human existence in the natural world) is the vehicle for the adoration and worship of God, where for members of the Royal Society—and the "moderns," more generally—that vehicle is instead the natural world itself: it is only by way of its "experimental" study (the "*profound* and thorow *Penetration* into her *Recesses*") that we can arrive first at knowledge and then, *consequently*, at the "Admiration" of God. Experimental philosophy, then, emerges as the mechanism that provides as one of its products the *mediated* return to God imagined on the model of natural knowledge. To the extent that it can stand synecdochically for science, the garden is figured not only as a point of destination—the result, one could say, of a scientific program that is simultaneously theological in nature—but also as the point of origin: the "parsly bed" (as the passage quoted earlier has it) "from whence we were dug." By these lights, the garden emerges as the machine that restores us to the presence of God—but only eventually and (what is more important) as a consequence of our experimental philosophy.

Evelyn's idyllic characterization of the garden—"[A]s no man can be very miserable that is master of a Garden here; So will no man ever be happy, who is not sure of a garden hereafter" (*EB*, 2)—emerges initially from a conventional understanding of the garden as paradise.[44] The term "paradise" was itself derived

---

[44] One can find similar sentiments in Abraham Cowley's poem "The Garden," for example (published in Evelyn's *Sylva*, 4th ed [1706: fol. ****]), in which the garden is nothing less than the nearest recreation of God's paradise:

> When God did Man to his own Likeness make,
> As much as Clay, tho' of the purest kind,
> By the great Potter's Art refin'd,
> Could the Divine Impression take;
> He thought it fit to place him, where
> A kind of Heav'n too did appear,
> As far as Earth could such a Likeness bear:

from the Persian *pairidaeza*, and introduced to Greek culture largely by way of the writings of Xenophon—a genealogy that draws the critical attention of Sir Thomas Browne.[45] Evelyn, writing at virtually the same moment Browne publishes *The Garden of Cyrus* (1658), offers an account of the beginnings of the social history of the garden:

> When Almighty God had exiled our Fore-fathers out of Paradise, the memorie of that delicious place was not yet so far obliterated, but that their early atempts sufficiently discover'd how unhappily they were to live without a Garden: And though the rest of the World were to them but a Wildernesse (and that God had destin'd them this employment for a sweete & most agreable purition of their Sinns) Adam instructed his Posteritie how to handle the Spade so dextrously, that in processe of tyme, men began, with the indulgence of heaven, to recover that by Arte and Industrie, which was before produced to them Spontaneously . . . [*EB*, 1]

And yet, Evelyn's conventional Christian garden history reveals something of a fracture: through their "Arte and Industrie," Adam and his descendents become so thoroughly proficient at their agricultural task—"to improve the Fruites of the Earth, to gratifie as well their Pleasures and contemplations"—that through the course of time these labors come to produce a natural perfection that, if it did exist in the natural world of Eden, had done so only *in potentia*:

> doubtlesse even in the most innocent state, though ther was no individual in itselfe imperfect, yet these perfections were to be discovered by Industry, & perhaps were not actualy existent & exerting their natures, & productions, till by his ingenuity they should afterwards be cultivated by such applications, marriages, combinations & experiments as his deepe knowledge in nature should prompt him to. [*EB*, 1][46]

This is an especially hopeful reading of the biblical curse laid upon Adam for his transgression in the garden. But, in keeping with the spirit of postlapserian labors guaranteed by virtue of the curse, Evelyn understands that the way toward the

> That Man no Happiness might want,
> Which Earth to her first Master could afford;
> He did a Garden for him plant,
> By the quick Hand of his Omnipotent Word.
> As the chief Help and Joy of human Life,
> He gave him the first Gift; first, ev'n before a Wife.

[45] Browne (1977: 327). Discussing the Persian conquest of Babylon—but the careful and deliberate preservation of its hanging gardens—Browne comments on paradise/*pairdaeza*:

> The *Persian* Gallants who destroyed this Monarchy, maintained their Botanicall bravery. Unto whom we owe the very name of Paradise: wherewith we meet not in Scripture before the time of *Solomon*, and conceived originally *Persian*. The word for that disputed Garden, expressing in the Hebrew no more then a Field enclosed, which from the same Root is content to derive a garden and a Buckler. [p. 327]

[46] Evelyn offers a similar sentiment in his *Kalendarium Hortense*:

> As Paradise (though of Gods own Phantasy) had not been Paradise longer then the Man was put into it, to Dress it and to keep it; so, nor will our Gardens (as neer as we can construe them to the resemblance of that blessed Abode) remain long in their perfection, unless they are also continually cultivated. [*Sylva* (1664)]

recovery of paradise as figured by the garden ("A place of all terrestriall enjoyments the most resembling *Heaven*") is essentially a matter of labor and struggle—renamed by Evelyn as "Arte" and "Industrie"—intended to construct order from the recalcitrant materiality of the natural world. The georgic scientific labor envisioned by Evelyn posits (and therefore can be said to create) a non-emblematical world that then becomes the object of his hortulan researches dedicated to the gradual but progressive accumulation of (practical) knowledge of that non-emblematical world. The frisson of Evelyn's garden-works (and the *Elysium Britannicum* in particular) is a function of this process of the artifaction of nature—a shudder that resonates most deeply and powerfully at precisely those moments when art and nature are brought into extreme proximity with one another. In one moment art is said to be secondary to nature, both later than and subservient to nature:

> What is in generall to be sayd, is, that it would be so contrived and set out, as that Art, though it contend with Nature; yet might by no meanes justle it out: There being nothing lesse taking, then an affected uniformity in greate & noble Gardens, where Variety were chiefly to be courted . . . For seing Nature dos in the universall oeconomy of things præceede Arte, and that Art is onely Natures ape, and dos nothing but by the power thereoff, as having first received all its principles from her Schoole (for thus the Physicks are ever before the Mathematicks) what can be more just and regular, then that she should also præside in the world about which our Gardiner (so much obliged to her) is perpetually conversant? [*EB*, 55][47]

And yet, in another moment, while describing "Artificiall Decorations" that may embellish the "natural" wonders and beauties of gardens—"Walls, Archtitecture, Particos, Terraces, Statues, Obelisks, Potts, Cascades, Fountaines, Basons, Pavilions, Aviaries, Coronary Gardens, Vineyards, [and] Walkes"—Evelyn asserts that any and all of these can be introduced into the garden "in their true & genuine places" without the gardener opening himself to the accusation of (artificial) "excesse" precisely because "Nature has already bin (as we may truly say) so Artificiall" (*EB*, 58).

Evelyn's blurring of the absolute distinction between art and nature underwrites his passion for including non-organic and non-living objects in his plan for the Royal Garden. Indeed, a significant portion of book two of the *Elysium* is dedicated to the careful articulation of the various artificial objects to be considered apt for inclusion in the garden: Chapter 9 is devoted to "Fountaines and [other] Hydropneumaticall Machines"; and Chapter 10 to "Rocks, grots, crypta's, Maunts, Precipices, Porticos, ventiducts." For his discussion in Chapter 11 of "Statues, Payntings, Columns, Dyals . . . ," Evelyn offers the following reflection upon the presence of non-living objects within the living world of the garden:

---

[47] Evelyn continues,

At no hand therefore let our Workman enforce his plot to any particular Phantsy, but, contrive rather how to apply to it the best shape that will agree with the nature of the Place; and studdy how even the most imperfect figure, may, by the Mysteries of Arte and fantsy, receive the most greacefull ornaments, and fittest for a Garden; yet in this proceedure, as not to inflame unnecessary expenses so, nor to Spoile it with another extreame, by a wastefull frugalitie: But when Nature will be more proper; then, to take leave of Art, & save the charges. [*EB*, 56]

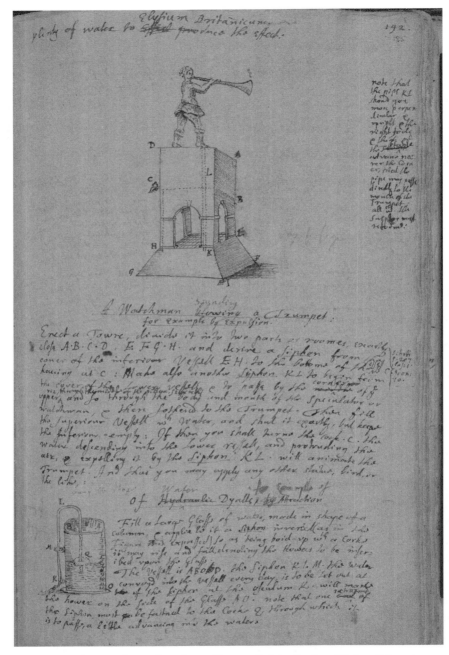

Figure 6.3 "A Watchman Sounding a Trumpet", John Evelyn, *Elysium Britannicum*

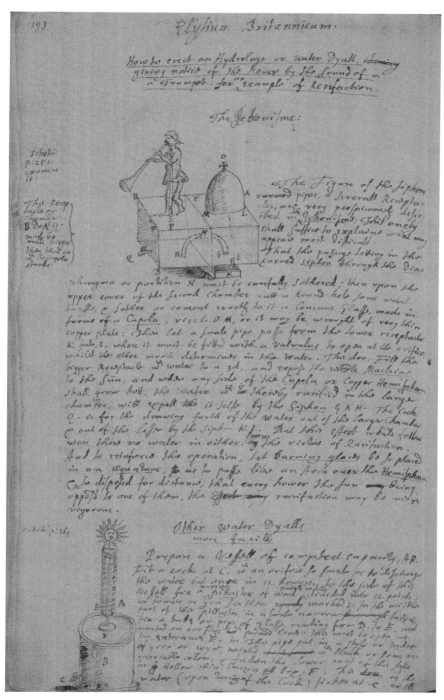

Figure 6.4 "Hydrologue", John Evelyn, *Elysium Britannicum*

As a Garden without Water hath no life, as depriv'd of its radicall humor, so without *Sculpture*, it has no action; for by this it is that we reppresent the figures of those great Heroes and Genious's that have so well deserv'd of Gardens, & so much celebrated by the Antients, affording an ornament not onely of exceeding pleasure to the eye, but to the intellect it selfe . . . [*EB*, 149]

And Chapter 12 is titled "Of artificial Echo's, Musick, & Hydraulick Motions." The many devices that feature in Evelyn's descriptions and in his technical drawings of the various "Motions" he would see populating the garden—the "Hydraulic Automat or Autophône Organ," the "Hydraulic Organ," and the "Hydraulic Automats" (*EB*, 176)—together constitute his most intensive (and fanciful) hybridization of the natural and the artificial (Figures 6.3 and 6.4).

Armed as he is by the belief that "a Garden hath of all other diversions, this prerogative alone, of gratifying the Senses virtuously" (*EB*, 167), Evelyn celebrates the virtues and varieties of production of "Artificiall musique" and uses his artificial and musical projections as a link and transition to consideration (in the following chapter) "Of Aviaries, Apiaries, Vivaries, Insects., etc.":

And where could this chapter have bin more appositely placed then after our discourse of Artificiall musiques to introduce the naturall. [*EB*, 200][48]

In *The Mangle of Practice: Time, agency and science*, Andrew Pickering describes scientists as "human agents in a field of material agency which they struggle to capture in machines."[49] While machinic culture is precisely the place where human and material agency enact a mutually constituting negotiation, and while its apparatuses are literal machines, I want to suggest that we can extend this idea of machinic culture to include science itself: science as a machine where we negotiate hybrid "emergent posthumanism" in which both human and material agency are at play, with neither encroaching upon or colonizing the other out of existence.[50]

Just as machinic captures of material agency mark stabilizing moments of posthuman science, so for Evelyn the confrontation between human agency and an interpretively resistant material agency are mutually stabilizing. In other words, in Evelyn's scientific practice, it is nature itself that has become the machine—the perfect hybrid that in fact stands between the terms of a series of only apparent binarisms: natural and artificial, nature and art, georgic and emblematical, labor and leisure, human and divine.

---

[48] As a way of explaining or illustrating his definition of artificial nature Evelyn turns immediately to two literary texts: Spenser's description of the Bower of Bliss (*The Faerie Queene*, 2.12) and, "if you will have it in poeticall prose, such a plot as the most accomplished *Sidny* makes Kalander to entertain the Prince Palladius" (*EB*, 58).

[49] Pickering (1995: 21). For Pickering's study, these machines are literal machines—the bubble chamber, for example, or high-energy particle accelerators—and they constitute the "surface of emergence" for his characterization of the "interactive stabilization" of "disciplined human agency and captured material agency" (p. 17).

[50] Pickering writes: "Resistances [i.e. moments when material reality confounds theoretical explanation] . . . are irrevocably impure, human/material hybrids, and this quality immediately entangles the emergence of material agency with human agency without, in any sense, reducing the former to the latter . . . [Also] human agency is itself emergently reconfigured in its engagement with material agency" (p. 54).

7

Time's Arrow

Nature, when interested in what we call life, appears to have a taste for the rococo.

C. P. Snow, *The Two Cultures: A second look*

To suggest, as I have in the preceding chapters, that the conventional separation of the literary-cultural from the scientific is itself an artificial act and does not reflect a natural or necessary distinction between them, is, at the same time, to offer (at least in broad terms) the possibility of the critique of a certain argument about origins and temporality—in short, an argument about the nature of history—that helps produce this traditional but finally very strange division of the kingdom in the first place. In other words, the edifice that emerges from the machinic scenario sketched in general terms in the introductory chapter, and then in the particular literary and scientific articulations addressed in Chapters 2 through 6, is always founded upon a certain theory of history that is thoroughly teleological in nature and that is therefore more a matter of ideology than anything else.

The critique of this theory of history would be no better, as it were, from the object of its analysis if it were itself to posit a point of origin, a moment to which we can turn in order better to understand the gradual unfolding of the allegedly natural separation of the literary and cultural from the scientific that represents, in many ways, the mature flowering of a theory of history that is more properly understood (as I will argue in the following pages) as a form of myth. It was in part for this reason that in the opening chapter of this study I turned to the *Gesta Grayorum* for a discussion of Francis Bacon, rather than to texts more conventionally construed as foundational to the new science—to *The Novum Organum*, say, or to *The Great Instauration*. The Gray's Inn Revels has never been identified as a foundational or originary moment of early modern science and for this reason (among others) it is a perfectly apt locus for a discussion aimed at displacing, as much as can be done, the search, explicit or implied, for something like the birth of science. This is a valuable undertaking not only because it is important to demonstrate the relative uselessness of a strictly etiological analytic *when the object of study is cultural in nature*, but also because to accept the search for origins is to accept the separation of the literary-cultural from the scientific that this study means to resist in the strongest terms possible.

With these brief remarks as prologue, and perhaps as cautionary, I would like to shift my focus in this final chapter from the texts and practices of the early modern period to a consideration of one especially enduring twentieth-century critical formulation of the questions that fall under the general rubric of the relationship between the literary-cultural and the scientific that I have been tracing throughout this book. I have in mind here C. P. Snow's remarkably durable book, *The Two Cultures and the Scientific Revolution* (1959) and his reconsideration of the two cultures controversy that ensued, *The Two Cultures: A second look* (1963).[1] This attention to Snow's two cultures (the phrase I will use to designate his books and, equally importantly, their persistent afterlife in debates about culture and science over the last fifty years) does constitute a change of immediate focus from the literary and scientific works of Shakespeare and Galileo and their times to consideration of some of the critical methods and historical assumptions that underpin contemporary debates. At the same time, I want to stress that there are nevertheless important through-lines of continuity that can be said to connect the foregoing discussion of early modernity and the present engagement with the two cultures. In this chapter, three of these continuities will be of special importance.

First, Snow's two cultures has been so highly influential that it stands for us today not merely as a single moment in the history of contemporary treatments of its theme, but rather as that analysis that has indeed served to structure the contemporary debate—from the identification of a *culture* of science, to the level of the vocabulary in which it has effectively memorialized the two competitors and the controversies that have emerged in their names.

Secondly, turning to Snow's two cultures provides me the opportunity to pick up an issue that surfaced in the preceding chapter's discussion of the idea—located, so I argued, in seventeenth-century garden practice, but increasingly pervasive across the period and across disciplines—that nature at times came to be understood as artificial. In that discussion, the figure of the garden, in conception, design, and in construction, served to embody this idea that redefined the nature of nature in ways (as I hope to outline in this chapter) that we are still negotiating today.

And thirdly, the following discussion will allow further consideration of a matter that has been at stake in all of the chapters so far, though it has not been addressed with strict and explicit focus until this point: the role that history plays in the critical, theoretical, literary, and scientific practices surveyed in this study.

In order to engage these continuities, it is necessary to address Snow's two cultures as he understood and articulated them. As texts now more often cited than read, and more often condemned than considered for what they actually try to argue, *The Two Cultures* and *A Second Look* remain pertinent to discussions of science and culture today, even if what one concludes is that Snow was, in some technical sense, "wrong" in his final analysis—though I am not prepared to make this determination. But this brings me to another important point about method

---

[1] Snow delivered his famous lecture at Cambridge University in May of 1959. It was subsequently published in numbers 12 and 13 of *Encounter* (June and July, 1959). The lecture was soon after published in book form with Cambridge University Press, also in 1959. Citations in this chapter to *The Two Cultures and the Scientific Revolution* refer to Snow's text as it appears in Snow (1998). This edition also prints Snow's subsequent reconsideration of the two cultures debate, *The Two Cultures: A second look* (originally published by Cambridge University Press in 1963); citations to this text refer also to this 1998 edition.

and how exactly this chapter will proceed. After having worked through the discussions of literary and scientific texts addressed so far, it should be clear by now that except in the most banal sense, the question of Right or Wrong is perfectly beside the point. To me, this means that what is at stake in reading C. P. Snow is not whether his argument can be said to be right or wrong (which of us is in a position to determine this?), but rather identifying the ways in which Snow's texts are constructed, how they function, and how precisely they mean anything at all. In other words, I propose to read Snow's two culture texts in the same ways in which we read *Hamlet*, or *The Winter's Tale*, or *Letters on Sunspots*, or *Devotions upon Emergent Occasions*, or early modern gardens and the texts that theorize them. On the level of method, a commitment to the theoretical aims of this study requires nothing less.

I

Snow begins *The Two Cultures* biographically, noting what he calls his own "unusual experience"—"By training I was a scientist: by vocation I was a writer" (*Two Cultures*, 1)—and the consequences of this "unusual experience" for his understanding of these two different "identities" and the cultural locations or sites where each is traditionally (and virtually inevitably) located:

> I have had, of course, intimate friends among both scientists and writers. It was through living among these groups and much more, I think, through moving regularly from one to the other and back again that I got occupied with the problem of what, long before I put it on paper, I christened to myself as the "two cultures." [*Two Cultures*, 2]

Lest his readers understand his point as merely biographical, Snow quickly broadens and, one can say, universalizes his sense of the two cultures: "No," he writes, "I intend something serious. I believe the intellectual life of the whole of western society is increasingly being split into two polar groups." Snow then states his fundamental thesis, to be elaborated through *The Two Cultures* (and which will be discussed in greater detail later in this chapter):

> Literary intellectuals at one pole—at the other scientists, and as the most representative, the physical scientists. Between the two a gulf of mutual incomprehension—sometimes (particularly among the young) hostility and dislike, but most of all lack of understanding. They have a curious distorted image of each other. Their attitudes are so different that, even on the level of emotion, they can't find much common ground. [*Two Cultures*, 4]

For all its remarkable staying power in the half-century since his Rede lecture at Cambridge, Snow's figuration of the two cultures should certainly not be thought of as a point of origin—either in general or for the following discussion. Snow can hardly be said to have inaugurated discussion of the relationship between the literary-cultural and the scientific, and still less this great divide itself. And yet, Snow's formulation has had a remarkably long-lived notoriety and, perhaps, even power. Evidence for this ongoing interest is evident in many dimensions and locations in today's world. We encounter it, certainly, in academic work, and we see it in popular print media.[2] In an essay entitled "Our two cultures" that appeared in the March 22, 2009 *New York Times Book Review,* to consider one instance, science journalist Peter Dizikes considers Snow and his legacy, and begins by noting that "Few literary phrases have had as enduing an afterlife as 'the two cultures,' coined by C. P. Snow to describe what he saw as a dangerous schism between science and literary life."[3]

The opening paragraphs of Dizikes's essay, in fact, contain what one quickly comes to notice as the battery of high-profile features of Snow's argument that appears over and over again in contemporary discussion of Snow's thesis, including Snow's double set of credentials as both a scientist and a novelist, Snow's central assertion that "the intellectual life of the whole of Western society is being split into two polar groups," F. R. Leavis's vitriolic condemnation of Snow ("as intellectually undistinguished as it is possible to be"). Dizikes's essay also quotes NASA's chief Michael Griffin's 2007 assertion that the two cultures is today "a bumper-sticker phrase" and, to even better effect, quotes Lawrence Summers's 2001 presidential inauguration address at Harvard: "We live in a society, and dare I say a university, where few would admit—and none would admit proudly—to not having read any plays by Shakespeare," even though, Summers adds, "it is all too common and all too acceptable not to know a gene from a chromosome."

Dizikes's reaction: "This is Snow for the DNA age." This is an observation that, while meant (as I take it) to be lightly comic, is less fanciful than may first appear, since the connection between the two cultures and the DNA age is in fact established by Snow himself precisely in his writings on the two cultures. At the same time, "Snow for the DNA age" signals something of a historical argument that not only links Snow of the 1950s to one of the nearly talismanic icons of modern scientific knowledge, but that also works to dissolve the time that separates Snow from us today. In other words, it is an historical argument that has the effect of collapsing history and the (linear) passage of time. As such, Snow for the DNA age comes to embody exactly the curious (and perhaps disturbing) fact of Snow's relevance today, even though his two-cultures argument is also firmly rooted in (for example) the ethos of 1950s modernization theory, or (to cite another context) the moment of decolonization. While I do not want to put more weight on this single article—however perceptive it may be—than it can fairly be asked to bear, it

---

[2] The literature on Snow and the two cultures controversy is vast; for an especially helpful recent contribution (and one that is particularly interested in the "science wars" that in many ways follow from the debate Snow initiated), see Labinger and Collins (2001). See also Spiller's (2004) discussion in her afterword ("Fiction and the Sokal hoax"), and Waugh (1999). See also Leith (2009). In his introduction to the collection of essays (1987), Levine begins by citing—in order only immediately to dismiss—Snow's *Two Cultures*: "'One Culture': the title seems to make a promise that this book will not keep. It obviously echoes Snow's 'two cultures,' by now a not very helpful cliché, and promises a unity we will not find." Then, citing an inadequacy that requires no discussion ("Snow's analysis was inadequate in ways that more critics than F. R. Leavis have noted"), Levine dismisses *The Two Cultures* altogether (p. 3).

[3] Dizikes (2009: 23).

is important to point out that all of this leads Dizikes to a question about history, cast in the form of progress:

> *The Two Cultures* actually embodies one of the deepest tensions in our ideas about progress. Snow, too, wants to believe the sheer force of science cannot be restrained, that it will change the world—for the better—without a heavy guiding hand. The Industrial Revolution, he writes, occurred "without anyone," including intellectuals, "noticing what was happening." But at the same time, he argues that 20th-century progress was being stymied by the indifference of poets and novelists. That's why he wrote *The Two Cultures*. So which is it? Is science an irrepressible agent of change, or does it need top-down direction?

Of course, science for Snow can both be an agent of (historical) change and at the same time stand in serious need of governmental organization and leadership. These two points—together with the teleological theory of history identified above—were on conspicuous display in a November 3, 2006 speech by Britain's Prime Minister Tony Blair given to the Royal Society and entitled "Britain's path to the future: Lit by the brilliant light of science." As even this title suggests, Blair's vision is founded upon an understanding of science and history—or, even, on science *as* history.

In the face of the familiar catalog of current economic transformations known collectively as "economic restructuring," Blair means to inspire hope by promoting the idea of "Britain's future as a 'knowledge' economy":

> For Britain, science will be as important to our economic future as stability. We have to be a magnet for scientific endeavour, attracting the best people, turning the knowledge into commercial enterprise, forming the collaborations and partnerships here, in Europe and across the world that keep us right at the frontiers that science is perpetually staking out.[4]

The potential obstacles barring the way to this imagined economy based upon commercialized science are many: the national education system needs to help young people become excited about science and want to embrace it; scientists and engineers need to be inspired to wed the scientific to the commercial; and Britain at large must mobilize all of its intelligence, innovation, and creativity. In fact, many of these potentialities are already within science itself; Blair points to one unnamed company (corporate secrets must remain secure) "run by engineers turned entrepreneurs" and its projects (including research on alternative energy for powering diesel engines on plant oil "with very much reduced levels of carbon emissions") and concludes: "In that small company are represented the many faces of science today: invention; practical application; commercial development and moral impulse."

But even given recent progress toward the realization of the new knowledge economy ("science is in many ways the secret success story of the Government"), much work remains to be done, especially in addressing the single most entrenched obstacle: "anti-science." This is work, Blair declares, that "won't be done by lofty

---

[4] Blair (2006, unpaginated). All subsequent quotations are from the text of this speech as it appears on this web page.

superiority but by engagement with the street, with science out there talking, debating, listening and educating." This need for "engagement with the street" in order to help move Britain into its bright future is one of long standing and has its own history; "These debates," Blair continues, "aren't new":

> In 1959, in Cambridge, C. P. Snow famously suggested that there exist in this country two separate cultures, all but unknown to each other, of the sciences and the arts. It was, he said, somehow culturally acceptable to be ignorant of the second law of thermodynamics in a way in which ignorance of Shakespeare was not.

Although Blair will concede that "[n]early half a century on, the sciences have become more specialized and popular understanding of its intricacies is, if anything, even worse," he nevertheless wants to mark Snow's paradigm as obsolete: "the idea of the two cultures seems completely out-dated." This notion is never explicitly argued or established in the text, even provisionally. Rather, the out-of-datedness seems instead to be merely a function of the passage of time, the defining feature of which is the inexorable progress of science. The result of this linked pair of forces, the passage of time and the inexorable progress of science, as Blair evidently judges it, is the disappearance of the two cultures: science has in effect devoured its opposite ("the arts") so that we are left with just *one* culture: "Science cannot any longer be detached from the society that houses it," Blair announces. "Its influence is too pervasive for that. Every area of policy today has a scientific aspect. Think of the big questions of our time—climate change, the spread of infectious diseases, water supply, biodiversity, terrorism. We will need to consult the scientists over every one."[5]

Rather than serving to distinguish this vision of new science and scientific society from the allegedly outdated version presented in *The Two Cultures*, however, many of Blair's attempts to promote the absolute value of commercial science and the knowledge economy serve only to suggest his relative closeness to major features of Snow's 1959 argument. In spite of his assertion of the outdatedness of Snow's two cultures argument, Blair's speech, perhaps in spite of itself, restates a number of the arguments in Snow's account and reaffirms a number of similar assumptions. Chief among these similarities are an uncritical celebration of science, especially the idea of the inevitability of scientific progress; and an uncritical dependence on a triumphalist theory of history. Let me take each in turn.

1. The celebration of science

Blair's one culture is saturated with new scientific objects: "MRI scanning, developed in this very building; the contraceptive pill; modern infertility treatment; ultrasound scan for unborn babies; unlocking the double-helix structure of DNA;

---

[5] The oddity in this list is certainly "terrorism," which most readers would perhaps not understand as a "big question" that necessarily avails itself of a scientific solution. Perhaps the inclusion of terrorism here is better understood as a reference to Prime Minister Blair's support of George W. Bush's so-called War on Terror, and as such is more obviously—and correctly—located squarely (if controversially) within the realm of the political.

keyhole surgery; placing fluoride in the water supply; the portable defibrillator; the Hepatitis B vaccine; strained quantum-well lasers which contain the information used in CDs, DVDs and the internet; DNA fingerprinting." The list continues: lab-on-a-chip technologies, in-situ diagnostics, nanotechnologies, smart materials, electronic cryptography, nuclear power, GM foods, MMR, stem-cell research, and so forth. Among the most pressing issues confronting science Blair identifies climate change and in support of its scientific resolution, the PM waxes poetic:

> The science of climate change is the moon landing of our day. This is idealism in a technical language. The scientists and the idealists will, once again, be the same people. The discoveries in the laboratory will be matters of life and death. Nothing could be more vital, nothing could be more exciting.

The celebration of science that characterizes Blair's perhaps utopic vision is evident throughout Snow's *Two Cultures*. We see a striking instance of this in the first of two curious—and perhaps telling and important—references to Shakespeare in Snow's text (I will turn to the second instance, referenced above by both Lawrence Summers and Tony Blair, later in this discussion). This reference occurs early in the book, almost immediately after Snow has posited the foundational notion of the "two polar groups" into which contemporary intellectual and practical life is in the process of dividing: literary intellectuals and scientists. But rather than standing as a celebration of Shakespeare, the reference is part of Snow's more general celebration of science. "Between the two polar groups," Snow writes, there exists only "a gulf of mutual incomprehension—sometimes (particularly among the young) hostility and dislike, but most of all lack of understanding" (*Two Cultures*, 4). Citing the literary intellectual's tendency "to think of scientists as brash and boastful" and seeking to present an illustration that characterizes this habit (and these two mutually uncomprehending cultures more generally), Snow tries to ventriloquise the "restricted and constrained" tone of literary culture characteristic of the figure of T. S. Eliot ("who just for these illustrations we can take as an archetypal figure") and whose attempt (as Snow describes it) to revive verse drama as a viable form for modern authors is characterized by little hope and the altogether modest ambition "that he would feel content if he and his co-workers could prepare the ground for a new Kyd or a new Greene." Snow concludes, "That is the tone, restricted and constrained, with which literary intellectuals are at home: it is the subdued voice of their culture" (*Two Cultures*, 4). To make the contrast explicit—and explicitly in favor of science over literary culture—Snow cites another voice, "a much louder voice, that of another archetypal figure, [Ernest] Rutherford, trumpeting: 'This is the heroic age of science! This is the Elizabethan age!'" Snow continues:

> Many of us heard that, and a good many other statements beside which that was mild; and we weren't left in any doubt whom Rutherford was casting for the role of Shakespeare. What is hard for the literary intellectuals to understand, imaginatively or intellectually, is that he was absolutely right. [*Two Cultures*, 4–5][6]

---

[6] Snow adds, "And compare 'this is the way the world ends, not with a bang but a whimper'—incidentally, one of the least likely scientific prophecies ever made—compare that with Rutherford's

Similarly, having chided those "self-impoverished" scientists for whom "traditional culture doesn't seem ... relevant" (*Two Cultures*, 13–4), Snow addresses what he sees as a certain chauvinism on the part of literary intellectuals when it comes to the question of culture—what it "is" and who can be said "to have" it. In the face of the attitudes of traditional culture, Snow is almost theatrically incredulous:

> But what about the other side? They are impoverished too—perhaps more seriously, because they are vainer about it. They still like to pretend that the traditional culture is the whole of "culture", as though the natural order didn't exist. As though the exploration of the natural order was of no interest either in its own value or its consequences. As though the scientific edifice of the physical world was not, in its intellectual depth, complexity and articulation, the most beautiful and wonderful collective work of the mind of man. Yet most non-scientists have no conception of that edifice at all. Even if they want to have it, they can't. It is rather as though, over an immense range of intellectual experience, a whole group was tone-deaf. Except that this tone-deafness doesn't come by nature, but by training, or rather the absence of training. [*Two Cultures*, 14] [7]

In both Blair's speech and Snow's *Two Cultures*, science is understood as an absolute good. And for both, science is also understood as salvational: the proper continued development of science will lead to Blair's "knowledge economy" that will have the double good of securing a viable economic future for Britain and at the same time re-establishing something like British scientific and technological hegemony in the twenty-first century as the inevitable outcome of nineteenth-century innovation. For Snow (and I will address this in more detail later in this discussion), science likewise represents the solution to national and global crises in education, in humanitarianism, and in morality. But even if these outcomes are imagined as unequivocal goods (and that is a debatable assertion), it is important to note that the route to these outcomes—the way, that is, for the continuation of the liberatory (and highly profitable) spread of science across the world—is secured through the workings of an uncritical faith in a certain theory of history. The model of an unproblematical and inevitable progress of science that founds much of the optimism evident in Blair's speech and Snow's book is itself the very hallmark of a utopian understanding of history as progressive and moral.

2. Triumphalist theory of history

For all of his forward-looking rhetoric, Blair's answer to current problems and the realization of the future's knowledge economy lie precisely in the past. In order to answer the present challenges the PM argues that Britain needs to nurture its

---

famous repartee, "Lucky fellow, Rutherford, always on the crest of the wave" Well, I made the wave, didn't I?" (*Two Cultures*, 5).

[7] Snow suggests that "while no one was looking" literary intellectuals "took to referring to themselves as 'intellectuals' as though there were no others." He then recalls an exchange ("some time in the 1930's") with the noted Oxford and Cambridge mathematician G. H. Hardy, who remarked to Snow "in mild puzzlement": "'Have you noticed how the word "intellectual" is used nowadays? There seems to be a new definition which certainly doesn't include Rutherford or Eddington or Dirac of Adrian or me. It does seem rather odd, don't y'know'" (*Two Cultures*, 4).

"capacity for ingenuity" and turn it "to practical, often commercial use." The guarantor of this eventuality is what Blair calls "Britain's industrial hegemony of the 19th Century," brought about by "the inventors of Lancashire: Arkwright's spinning jenny, Kay's flying shuttle, Crompton's spinning mule." Later in the speech Blair in fact will assert that the UK is "very well placed" to achieve the science-based knowledge economy and the reason for this is that it has "rediscovered the secret of the first industrial revolution. We have brought scientific discovery and entrepreneurial activity together."

And Snow, for his part, because the scientific revolution follows *as an historical consequence* from the industrial revolution, it is in effect the contemporary form of its continuation and gradual unfolding across the last two centuries. Together these revolutions constituted a transformation of such colossal proportion and significance as to be rivaled only by the invention of agriculture. "In fact," Snow writes, "those two revolutions, the agricultural and the industrial–scientific, are the only qualitative changes in social living that men have ever known" (*Two Cultures*, 23). But the matter of the *meaning* of this extended revolution remains elusive: "do we understand how they have happened? Have we begun to comprehend even the old industrial revolution? Much less the new scientific revolution in which we stand? There never was anything more necessary to comprehend" (*Two Cultures*, 28).

In order to clarify the nature of his questions about historical causes and our contemporary understanding, Snow offers more precise definitions of the two phases of the great revolution. "By the industrial revolution, I mean the gradual use of machines, the employment of men and women in factories, the change in this country from a population mainly of agricultural labourers to a population mainly engaged in making things in factories and distributing them when they were made" (*Two Cultures*, 29). The natural emergence of the scientific revolution follows from the fundamental change represented by the birth of industry:

> Out of it grew another change, closely related to the first, but far more deeply scientific, far quicker, and probably far more prodigious in its result. This change comes from the application of real science to industry, no longer hit and miss, no longer the ideas of odd "inventors", but the real stuff. [*Two Cultures*, 29][8]

This sounds rather like Tony Blair's impassioned call for commercialized science. Both versions are underwritten by the unproblematical celebration of "Britain's industrial hegemony of the 19th century" (as Blair has it) and both predict the economic, political, and even the *moral* rescue of British and Western national interests. This is in itself a remarkable fact, given the very many differences that distinguish Blair's moment from Snow's. And yet, even given these differences, both men seem to operate according to an unstated (and perhaps unselfconscious)

---

[8] Snow is concerned to date this second phase: "when atomic particles were first made industrial use of." He continues:

> I believe the industrial society of electronics, atomic energy, automation, is in cardinal respects different in kind from any that has gone before, and will change the world much more. It is this transformation that, in my view, is entitled to the name of "scientific revolution." [*Two Cultures*, 30]

Whiggish notion of history. And while both Blair and Snow can be said to construct their respective worlds at least in part upon an understanding of the two-cultures debate, the solutions they imagine are nevertheless distinct: Blair's proposal (as I will argue presently) is political in nature, while Snow's is humanitarian and educational.

For Blair the entire question of science, especially its relation to an imagined "knowledge economy" and Britain's future, is fundamentally a political one, with both a political history and—more importantly—a political solution. Citing a worrisome history ("ever since the 1780s...") of a certain slowness "to turn its technological and scientific creativity to economic purpose," Blair recounts his government's efforts to change this tendency: an increase in the science budget from £1.3 to £3.5 billion, enhanced investment in university science infrastructure, the establishment of the Ten-Year Science and Innovation Investment Framework ("We have been the first government to set out a long-term vision in science"), tax relief on research and development, the UK Stem-Cell Initiative, among other initiatives and programs. "[T]he basis for world-class science," the PM declares, "is in place."

A major challenge remains, however, and since that challenge is essentially political in nature, the response to it is, as well. "Government must show leadership and courage in standing up for science and rejecting an irrational debate around it." The great obstacle is "anti-science," which takes two linked forms: activism, on the one hand, and intellectual skepticism (or even resistance), on the other. On the former, Blair points to "a powerful and vocal lobby, with access to all the media channels and an interest in polarising the argument [that] frames the debate." Then, in polarizing language of his own (the inevitable by-product of oppositional rhetoric), Blair spies the enemy: "The anti-science brigade threatens our progress and our prosperity. We need political and science leadership that stands up to them." Resisting this powerful special interest is a difficult task. Blair offers a specific instance of the "classic...struggle between short term politics and long term public good":

> If we hadn't taken on the animal rights extremists, we might well have lost essential scientific research to Britain with incalculable economic damage to the country to say nothing of the value of the research in the treatment of disease. But in a sense, that was an argument, because of the violence, we could easily enough win.

The implication here is clear: other battles may not be so easily won—or won at all. Citing debates over nuclear power, GM foods and drugs, and stem-cell research as prominent examples, Blair warns of the still greater difficulty that lies in "the genuine areas of intellectual controversy." The warning is rendered still more ominous: the "misconceptions, often born of the most outrageous distortion of fact by campaigners, who in accusing others of a lack of scruple show precious little of it themselves, can be so pervasive. They so easily take hold." This, in turn, leads to something of a peroration that culminates with a vision of a scientific democracy:

Combating them takes the world of science to engage fully, clearly and in simple language with the world outside it. We need scientists willing and able to explain, to reason, to give the scientific facts not by arrogant assertion but by patience and also accurately reflecting where science is fact and where it is still conjecture. Britain as a whole must become a scientifically literate society. This is not simply to grow the next generation of scientists but also to condition all of us to a reasoned understanding of what science can do for us; to dispel the myths; calm the scares; let us make our moral judgements, at least partially, on the facts.

As one example of how this new science culture works, Blair offers the case of BSE—Bovine Spongiform Encephalopathy—as a "template of how to conduct a rational conversation about science." Noting that the first step is "to ensure we hear scientific truth told to power in government," Blair outlines the two phases his government's procedures followed. The first phase:

> Following the Phillips report into BSE in 2000, we appointed Chief Scientific Advisers in all major departments. We opened up scientific advisory committees to greater scrutiny. We created an independent Food Standards Agency to ensure we transfer the best scientific knowledge to matters of obvious sensitivity to the public.

The second phase: engage the public at an early stage. At the same time, there are two further imperatives: "We then need always to be clear about how the benefits accrue to individuals," and "we must then be honest about the risks."

With these "conditions" in place, Blair argues, "a genuinely open, rational dialogue is possible." But in spite of the notion of an actual debate, the outcome the Prime Minister clearly imagines is predetermined: science will lead and, if it wishes to succeed, society must follow. The only matter, then, is how government will facilitate this forward progress. In other words, the dialogue about science is no dialogue at all. This is an outcome underwritten by a number of assumptions: that when it speaks to "power," science knows what it is saying; that science is monological; and that a fullness of knowledge will lead necessarily to *moral* moral judgments.

What is especially striking about the Prime Minister's speech is the great extent to which his vision of a knowledge economy can be realized only by means of a rhetoric of erasure that has the effect, and the goal, of establishing a single culture. This vision of a culture constructed on the active belief (which is to say a *politics*) based on a vision of the ascendency of science requires that it consume everything in its path, including what Blair identifies as the "anti-science brigade." But this rhetoric of erasure itself is leveraged on the absolute and necessary presence of the second culture it targets for elimination: science can only be imagined as speaking to power if power is first imagined as somehow separate from science; science can also only be imagined as monological if politics is imagined as dialogical; and one can contemplate *moral* moral judgments within a plurality of cultures in which such a differential choice is available in the first place. Likewise, Blair can imagine engaging a public if that public is itself construed as somehow outside the culture of science, since (presumably) individuals within the culture of science do not stand in such need of active—and persuasive—engagement in pursuit of social and political action undertaken in something like an evangelical effort on behalf of the culture of science.

I will take up the matter of Snow's humanitarian–moral solution in the next section of this chapter, but for the moment it is important to note some of its principal features, especially those that contrast with Blair's. Most significantly, Snow's understanding of both the problem of the two cultures and his solution to it are not centered on Britain—or the West, for that matter. Rather, Snow's argument, which I will describe as historical in nature, is grounded in his understanding of the plight of the Third World and what we today call developing nations (in Snow's era, and in Snow's account, these latter are Russia, China, and India). The great problem, on Snow's account, is the disturbing and perhaps catastrophic emergence of a global culture of have's and have-not's, the only solution to which, he argues, lies in the spread of science and technology from the West to the East. In other words, Snow's account is grounded in what he argues are the humanitarian and moral obligations that face Britain and the United States; his solution, accordingly, is less about internal or national politics, than in something like our global fraternal obligation to our fellow man in an ever-increasingly global social setting.

If the politician Blair offers a political solution, and if Snow as humanitarian and educator (as least he is so in his role as public intellectual from which he offers *The Two Cultures*) offers a humanitarian and educational solution, what should we expect if we consider (even if only briefly) what a scientist would offer as a solution? Are scientists likely to subscribe to the notion of the two cultures? Have they addressed this debate as pertinent today (even in spite of Blair's assertion of the outdatedness of the two cultures)? I would like to offer a brief consideration of one prominent scientist—one who has also assumed the mantle of public intellectual— for whom Snow's argument remains, perhaps worrisomely, relevant. Less than a decade before the Prime Minister's declaration of the essentially obsolete nature of Snow's two cultures (and nearly fifty years after Snow's Rede Lecture), the eminent biologist Edward O. Wilson published *Consilience: The unity of knowledge*, a work very much addressed to the what Wilson takes as the *fact* of the two cultures. Wilson grants that for "most scholars the two domains, commonly called the scientific and literary cultures, still have a look of permanence about them."[9] But the separation between these two cultures is not absolute:

> From Apollonian law to Dionysian spirit, prose to poetry, left cortical hemisphere to right, the line between the two domains can be easily crossed back and forth, but no one knows how to translate the tongue of one into that of the other. Should we even try? I believe so, and for the best of reasons: The goal is both important and attainable. The time has come to reassess the boundary. [p. 125]

Although Wilson does not expect universal agreement on his assessment of the need to reassess the boundary, he believes that "few can deny that the division between the two cultures is a perennial source of misunderstanding and conflict"

---

[9] Wilson (1998: 125).

(p. 125). Wilson then offers a quotation from what he calls Snow's "defining 1959 essay *The Two Cultures and the Scientific Revolution*:" "This polarisation is sheer loss to us all. To us as people, and to our society. It is at the same time practical and intellectual and creative loss" (*Two Cultures*, 11). For Wilson, the "root cause of the problem" is "as obvious today as it was when Snow ruminated on it at Christ College high table: the overspecialization of the educated elite" (p. 126). As I will argue below, this assessment differs significantly from Snow's sense of the "root cause" (Wilson's term) of the divide between the two cultures. Indeed, Snow not only declines to identify overspecialization as the fundamental cause; his argument shifts across the essay in such a way as to replace the scientific and the literary/ traditional as his key binary terms with another pair: rich and poor. But for Wilson, whose argument is not, in the end, a cultural one, the "root cause" has in effect a root cure: gene–culture coevolution. Wilson writes:

> There is only one way to unite the great branches of learning and end the culture wars. It is to view the boundary between the scientific and literary cultures not as a territorial line but as a broad and mostly unexplored terrain awaiting cooperative entry from both sides. The misunderstandings arise from ignorance of terrain, not from a fundamental difference in mentality. The two cultures share the following challenge. We know that virtually all of human behavior is transmitted by culture. We also know that biology has an important effect on the origin of culture and its transmission. The question remaining is how biology and culture interact, and in particular how they interact across all societies to create the commonalities of human nature. What, in the final analysis, joins the deep, mostly genetic history of the species as a whole to the more recent cultural histories of its far-flung societies? [p. 126][10]

Even as it is not surprising that Tony Blair's solution to the two cultures is political, or that Snow's is humanitarian and educational, so it is not surprising to learn that Wilson's is biological. What is interesting and worth highlighting is that even the scientific solution is itself at the same time an historical solution. For what else is the appeal to evolution but an implicit appeal to history? And even as Blair or Snow imagine the solution to the two cultures problem as resolvable by the means of the working out of history, so, too, does Wilson's gene–culture coevolution mark both a faith in and an appeal to history through the mechanism of the double helix (to which I will return later in this chapter).

If what I have suggested thus far is correct, then when our three writers consider the two cultures and pose solutions they are actually also constructing arguments about time and history. At the same time, their arguments about the nature of the relationship between the two cultures—between the scientific, on the one hand, and the literary or the traditional culture, on the other—remain pertinent, even fifty years beyond Snow's Cambridge lecture, and even in the aftermath of the so-called science wars. It is my sense that the enduring nature of this question is

---

[10] Wilson glosses the term: "In essence, the conception observes, first, that to genetic evolution the human lineage has added the parallel track of cultural evolution, and, second, that the two forms of evolution are linked" (p. 127). Wilson, together with Charles J. Lumsden, introduced the term/ concept of gene–culture coevolution (1981) and (1983).

itself something worth consideration. Why has the two-cultures debate endured? Has the debate moved forward in any sense, or are the sides as "polar" as Snow claimed in 1959? These, and similar questions, have been addressed with a sometimes astonishing rigor since Snow's lecture and show no obvious signs of slowing or abating. But that is an argument I do not wish to enter—at least not directly. For it seems to me that there is another question, a *prior* question (as it were) about the debate that has remained more or less obscured and overlooked in the debate. It is this question I wish to pursue here: is the scientific culture-literary culture (or, as Tony Blair casts it helpfully, the science-arts) debate best understood as a new or "modern" form of the ancient debate over the relationship between nature and art?

This is the question I seek to address in the following pages. In order to begin that project, I would like to return to a touchstone moment in Snow's *Two Cultures*: the appearance of Shakespeare. Snow's second reference to Shakespeare (as Blair notes in his speech before the Royal Society) occurs when he famously juxtaposes an assumed familiarity with Shakespeare's works to an assumed unfamiliarity with the second law of thermodynamics:

> A good many times I have been present at gatherings of people who, by the standards of traditional culture, are thought highly educated and who have with considerable gusto been expressing their incredulity at the illiteracy of scientists. Once or twice I have been provoked and have asked the company how many of them could describe the Second Law of Thermodynamics. The response was cold: it was also negative. Yet I was asking something which is about the scientific equivalent of: *Have you read a work of Shakespeare's?* [*Two Cultures*, 14–5][11]

But why does Snow invoke this particular law? Or, rather: how does the second law function in Snow's account? Evidently, the second law stands as the equivalent for a scientist to a work by Shakespeare for a literary intellectual: both are fundamental and both stand as the foundation upon which two cultures are constructed (an argument certainly open to vigorous debate). At the same time, both the second law and a work by Shakespeare emerge from the logic of Snow's argument as facts of nature: they stand as two instances of the given in our two cultures, even if they are not communicable across the cultural divide. But one can equally easily—or perhaps even more easily—understand a work by Shakespeare and the second law of thermodynamics as not only not related as (in effect) homologies, but rather as functional opposites. For all of its beauty, the second law as such is not understood by the scientist as a function of culture; it is resolutely not constructed, but rather stands as the given upon which something like the "natural order" Snow mentions is itself established. For its part, a work by Shakespeare may indeed strike a certain kind of traditional literary reader as foundational, but it is no more a product of nature than the notion of increasing entropy is a product of language.

---

[11] Snow continues,

I now believe that if I had asked an even simpler question—such as, What do you mean by mass, or acceleration, which is the scientific equivalent of saying, *Can you read?*—not more than one in ten of the highly educated would have felt that I was speaking the same language. [*Two Cultures*, 15]

So the question stands, then: what enables (or requires) Snow to link the second law and Shakespeare?

In response to this question, I would like to offer a discussion of crystallography and Snow's own comments on his use of thermodynamics as a scientific equivalent to literary culture's Shakespeare, as represented in *The Two Cultures: A second look*, Snow's 1964 essay on the two cultures and the controversy his earlier essay seems to have ignited.

II

The 1994 C. P. Snow Lecture at Christ College, Cambridge was delivered by Roy Porter and the resulting essay, "The two cultures reconsidered," was published in 1994 in *The Cambridge Review* and then again, in adapted form, in 1996 in *boundary 2*. In his essay, Porter notes how curious Snow's invocation of the second law is and would like, he writes, "to float the suggestion that the second law of thermodynamics was not mentioned lightly" (p. 2).[12] On Porter's account, the mention of the second law is "a reference back to, a recollection of, a passage in a work by J. D. Bernal, who became friends with Snow after returning to Cambridge in 1927" (p. 2). For Porter, Snow's reference to the second law looks back (in part) to Bernal's *The World, the Flesh, and the Devil* (1929) in which Bernal set out to describe the defeat by science of what he identified as "the three enemies of the Rational Soul." Porter summarizes:

> Biology would correct the shortcomings of the human body, psychology would regulate man's "desire and fears, his imagination and stupidities," while physics, the greatest of them all, would tame "the massive, unintelligent forces of nature," endowing mankind with tremendous new powers—though only up to a point, Bernal was forced to concede, on account of the limits set by the second law of thermodynamics; yet even when faced by entropy, concluded the communist crystallographer, "intelligent organization" would enable mankind in fair measure to frustrate the inevitable. [p. 3]

As this suggests, Bernal's vision of science is a strikingly utopic one and as such would certainly be attractive to Snow (and even to Tony Blair, for that matter). For Porter, however, the idea—or, perhaps, the *figuration*—of Snow's two cultures functions in a curious fashion, identifying (as it may do) the work of Bernal without an explicit reference to Bernal or his work appearing in Snow's study. Indeed, the reference works (if it works) without even an implicit reference. Instead, as Porter has it, Snow's two cultures is—at least in part—a response to an absent stimulus, or an absent presence. For Porter, then, the invocation of the second law is in fact a reference to an entire "historical narration" that Snow declines to engage: the "mention of the second law of thermodynamics may have been [Snow's] way of making oblique acknowledgment that he was picking at the threads of an earlier scholarly tapestry" (p. 3).

---

[12] Porter (1996: 2).

Porter's descriptor is "oblique." But the connection to Bernal in Snow's essay is more complicated. The two items of information that Porter offers without comment or discussion but that are of primary importance to the present discussion is that Bernal was a communist and that Bernal was a crystallographer. For the two cultures, these pieces of information are linked.

Snow was not a communist; nor was he a crystallographer. His generally leftist politics, however, are clearly on display in *The Two Cultures*, as is a certain concern with the contemporary societal equivalent of the physical and chemical features of crystallization: the ossification of the British class system and the entire range of the social structures of privilege. Snow wastes no time in establishing his own "poor" origins. At the outset of the essay Snow comments on the social and economic "circumstances" of his early life and suggests that "[a]nyone with similar experience" in life would come to the same observations and conclusions about the two cultures as he had: "By training I was a scientist; by vocation I was a writer. That was all. It was a piece of luck, if you like, that arose through coming from a poor home" (*Two Cultures*, 1). And so it follows that part of what is so compelling about science for Snow is that it is a great leveler. When describing scientific culture as a culture—"scientific culture really is a culture, not only in an intellectual but also in an anthropological sense" (*Two Cultures*, 9)—Snow cites "common attitudes, common standards and patterns of behaviour, common approaches and assumptions" that members of the scientific culture share. "This goes surprisingly wide and deep," Snow writes. "It cuts across other mental patterns, such as those of religion or politics or class" (*Two Cultures*, 9). For Snow, even though more scientists tend to be "unbelievers" than do other intellectuals, even as more scientists "are on the Left in open politics" and "considerably more scientists in this country and probably in the U.S. come from poor families," these differences are of little consequence since in "their working, and in much of their emotional life, their attitudes are closer to other scientists than to non-scientists who in religion or politics or class have the same labels as themselves" (*Two Cultures*, 10). And this holds true, he writes, for conservatives and for liberals:

> That was as true for the conservatives J. J. Thomson and Lindemann as of the radicals Einstein or Blackett: as true of the Christian A. H. Compton as of the materialist Bernal: of the aristocrats de Broglie or Russell as of the proletarian Faraday: of those born rich, like Thomas Merton or Victor Rothschild, as of Rutherford, who was the son of an odd-job handyman. Without thinking about it, they respond alike. That is what a culture means. [*Two Cultures*, 10]

For Snow, science is a powerful antidote to what we might call the special interests of wealth and privilege. This is a major theme of Snow's work on the two cultures and helps situate what I earlier called his humanitarian and educational program for global social equalization through the offices of science and technology.

For Snow, the two cultures is a Western problem, though the "cultural divide" is, as he sees it, "at its sharpest" in England. There are two related causes: "our fanatical belief in educational specialisation" (which Snow will address at length in the essay) and, larger and more significantly, "our tendency to let our social forms crystallise," a tendency Snow believes is intensifying:

> This tendency appears to get stronger, not weaker, the more we iron out economic inequalities: and this is especially true in education. It means that once anything like a cultural divide gets established, all the social forces operate to make it not less rigid, but more. [*Two Cultures*, 17]

As this passage suggests—and as much of the subsequent work of his essay will manifest—Snow's argument is an historical one: as time passes and "social forces" continue to crystallize the cultural divide, the opportunities for communication across that divide diminish: "Thirty years ago the cultures had long ceased to speak to each other: but at least they managed a kind of frozen smile across the gulf. Now the politeness has gone, and they just make faces" (*Two Cultures*, 17).[13]

For Snow, the only answer to the current crisis is through educational reform. This notion leads to an extended and comparative discussion of British—and American and Russian—educational systems, a discussion, it is well to point out, that constitutes the bulk of the essay. Other nations have taken notice of the dangers of overspecialization in education and are "seized of the problem." But, Snow asks, "Are we? Have we crystallised so far that we are no longer flexible at all?" (*Two Cultures*, 19).

To illustrate the set of problems that have resulted from a history of social and educational crystallization that has led to the present crisis—"All the lessons of our educational history suggest we are only capable of increasing specialisation, not decreasing it"—Snow offers the only example "in the whole of English educational history, where our pursuit of specialised mental exercise was resisted with success" (*Two Cultures*, 19): the abolition of the "old order-of-merit" in the Mathematical Tripos at Cambridge. "For over a hundred years," Snow writes, "the nature of the Tripos had been crystallising." The Tripos examination was "intensely competitive and intensely difficult" and a "whole apparatus of coaching" had developed around it. "Men of the quality of Hardy, Littlewood, Russell, Eddington, Jeans, Keynes, went in for two or three years' training" in order to prepare. Any commentary on the entire system was colored by an intense pride ("similar to that which almost anyone in England always has for our existing educational institutions, whatever they happen to be"): the test insured the highest standards, fairly tested for sheer merit, and was indeed "the only seriously objective test in the world" (*Two Cultures*, 20).[14] The system seemed immutable because of its very nature. "In every respect

---

[13] Snow continues:

It is not only that the young scientists now feel that they are part of a culture on the rise while the other is in retreat. It is also, to be brutal, that the young scientists know that with an indifference degree they'll get a comfortable job, while their contemporaries and counterparts in English or History will be lucky to earn 60 per cent as much. [*Two Cultures*, 17–8]

[14] Snow draws the similarities to the contemporary system of scholarship examinations; the arguments in support of the old Tripos system "were almost exactly those which are used today with precisely the same passionate sincerity if anyone suggests that the scholarship examinations might conceivably not be immune from change" (*Two Cultures*, 20).

but one, in fact, the old Mathematical Tripos seemed perfect." And then Snow delivers the bad news:

> The one exception, however, appeared to some to be rather important. It was simply—so the young creative mathematicians, such as Hardy and Littlewood, kept saying—that the training had no intellectual merit at all. They went a little further, and said that the Tripos had killed serious mathematics in England stone dead for a hundred years. [*Two Cultures*, 20–1]

In the end, the "creative mathematicians" won and the system was changed. But Snow is concerned that "Cambridge was a good deal more flexible between 1850 and 1914 than it has been in our time," and he wonders if "we had the old Mathematical Tripos firmly planted among us, should we have ever managed to abolish it?" (*Two Cultures*, 21).

The answer, it would seem, is No, and the reason—or at least a significant part of the reason—is the degree of crystallization that has resulted in something like social, educational, and perhaps even moral stagnation. Snow's discussion now turns more obviously toward an analysis of history and he begins with the industrial revolution and culture's response to it. Apart from scientific culture, Snow argues, "the rest of western intellectuals have never tried, wanted, or been able to understand the industrial revolution, much less accept it. Intellectuals, in particular literary intellectuals, are natural Luddites." This fact, as Snow sees it, helps to account for "our present degree of crystallisation" (*Two Cultures*, 22). Even though the industrial revolution, together with the discovery of agriculture, constituted "the only qualitative changes in social living that men have ever known," intellectuals and traditional culture simply "didn't notice":

> or when it did notice, didn't like what it saw. Not that the traditional culture wasn't doing extremely well out of the revolution; the English educational institutions took their slice of the English nineteenth-century wealth, and perversely, it helped crystallise them in the forms we know.
>   Almost none of the talent, almost none of the imaginative energy, went back into the revolution which was producing wealth. The traditional culture became more abstracted from it as it became more wealthy, trained its young men for administration, for the Indian Empire, for the purpose of perpetuating the culture itself, but never in any circumstances to equip them to understand the revolution or take part in it. [*Two Cultures*, 23][15]

This error is virtually universal in the West. And writers (as exemplars of traditional culture) were as guilty in this willed blindness: "Plenty of them shuddered away, as though the right course for a man of feeling was to contract out; some, like Ruskin and William Morris and Thoreau and Emerson and Lawrence, tried various kinds of fancies which were not in effect more than screams of horror" (*Two Cultures*, 25). And this leads to some of Snow's most strident language in support of the idea

---

[15] After identifying the historical nature of the causes of the current cultural crisis (and exonerating his countrymen "now living"), Snow writes: "If our ancestors had invested talent in the industrial revolution instead of the Indian Empire, we might be more soundly based now. But they didn't" (*Two Cultures*, 39).

that "Industrialisation is the only hope of the poor" and in condemnation of traditional culture's crystallization:

> It is all very well for us, sitting pretty, to think that material standards of living don't matter all that much. It is all very well for one, as a personal choice, to reject industrialisation—do a modern Walden, if you like, and if you go without much food, see most of your children die in infancy, despise the comforts of literacy, accept twenty years off your own life, then I respect you for the strength of your aesthetic revulsion. But I don't respect you in the slightest if, even passively, you try to impose the same choice on others who are not free to choose. In fact, we know what their choice would be. For, with singular unanimity, in any country where they have had the chance, the poor have walked off the land into the factories as fast as the factories could take them. [*Two Cultures*, 25–6])

What becomes clear through the middle portions of *The Two Cultures* is that the chief good of industry and science lies in their potential for positive social change. The problem—the challenge, as Snow sees it—is that this change is resisted by the forces of crystallization. This includes specific changes in, for example, the nature of the work done by the laboring classes (from agricultural work to industrial production and distribution)—a change, Snow argues, that "crept on us unawares, untouched by academics, hated by Luddites, practical Luddites and intellectual ones" connected "with many of the attitudes to science and aesthetics which have crystallised among us" (*Two Cultures*, 29). But it also requires massive changes in Western educational philosophy and practices. And it also requires the rich nations of the West extending the helping hand of practical science and technology to the poor nations of the world.[16] For Snow, all of this is understood as a matter of history: the legacy of the crystallization and the potential for change in the future. "To say, we have to educate ourselves or watch a steep decline in our own lifetime is about right" (*Two Cultures*, 39). Snow registers his deep concern, born of the lessons he has read from the past:

> We can't do it, I am now convinced, without breaking the existing pattern. I know how difficult this is. It goes against the emotional grain of nearly all of us. In many ways, it goes against my own, standing uneasily with one foot in a dead or dying world and the other in a world that at all costs we must see born. I wish I could be certain that we shall have the courage of what our minds tell us. [*Two Cultures*, 40]

And then Snow admits to being "saddened" by what he calls the "historical myth" of the Venetian Republic "in their last half-century":

> Like us, they had once been fabulously lucky. They had become rich, as we did, by accident. They had acquired immense political skill, just as we have. A good many of them were tough-minded, realistic, patriotic men. They knew, just as clearly as we

---

[16] The forms that such help would need to take, Snow argues, would be first, capital, and after that, "men":

> The second requirement, after capital, as important as capital, is men. That is, trained scientists and engineers adaptable enough to devote themselves to a foreign country's industrialisation for at least ten year out of their lives. [*Two Cultures*, 47]

know, that the current of history had begun to flow against them. Many of them gave
their minds to working out ways to keep going. It would have meant breaking the
pattern into which they had crystallised. They were fond of the pattern, just as we are
fond of ours. They never found the will to break it. [*Two Cultures*, 40]

All of this historical reasoning has led Snow, as he moves into the final chapter of his
essay, to something of a revelation, almost in spite of the argument he has offered to
this point. Troubled as he is by the thought that "the Venetian shadow falls over the
entire West" (*Two Cultures*, 41), he nevertheless declares that the historical argument (which I am tempted to call an *entropic* historical argument, as I will discuss
below) "isn't the main issue of the scientific revolution." Instead, the main issue is
"that the people in the industrialised countries are getting richer, and those in the
non-industrialised countries are at best standing still: so that the gap between the
industrialised countries and the rest is widening every day. On the world scale this is
the gap between rich and poor" (*Two Cultures*, 41). With glances as far back as
"Neolithic times," Snow argues that life "for the overwhelming majority of man-
kind has always been nasty, brutish and short. It is so in the poor countries still"
(*Two Cultures*, 42).

By this point in his essay, one can say that the two cultures are no longer the
scientific and the literary or the traditional, but rather the rich and the poor. It is by
means of this "new" understanding of the cultural divide that Snow seeks to
leverage his humanitarian arguments for social change achieved through the open
spread of practical science and technology in a movement from West to East, rich
to poor:

> This disparity between the rich and poor has been noticed. It has been noticed, most
> acutely and not unnaturally, by the poor. Just because they have noticed it, it won't last
> for long. Whatever else in the world we know survives to the year 2000, that won't.
> Once the trick of getting rich is known, as it now is, the world can't survive half rich
> and half poor. It's just not on. [*Two Cultures*, 42]

Snow closes *The Two Cultures* with a prediction: either the West will resolve the
problem of the cultural gap between the rich and the poor, or the "Communist
countries will in time" (*Two Cultures*, 50). This would represent not only the moral
failure of the West, but also the end of history as we have been pleased to
understand it:

> For, though I don't know how we can do what we need to do, or whether we shall do
> anything at all, I do know this: that, if we don't do it, the Communist countries will in
> time. They will do it at great cost to themselves and others, but they will do it. If that is
> how it turns out, we shall have failed, both practically and morally. At best, the West
> will become an *enclave* in a different world—and this country will be the *enclave* of an
> *enclave*. Are we resigning ourselves to that? History is merciless to failure. In any case,
> if that happened, we shall not be writing the history. [*Two Cultures*, 50]

To the considerable extent, then, that *The Two Cultures* is itself—and perhaps
unknown to itself, or, even, *in spite* of itself—a work of what we can call social
crystallography, it is at the same time necessarily a study of history in which social

change is seen to erupt in the very face of structures of privilege that are not only already in place, but that are further solidified by the *two* cultural gaps Snow seeks to understand: between science and literary-traditional culture, on the one hand, and between the global cultures of rich and poor, on the other. But, as I have tried to indicate, Snow's text is an unstable one: its target shifts from the first pair of cultures to the second, even as the precise relationship between them remains allusive. Moreover, as an argument about history, which is the actual nature of *The Two Cultures*, Snow's discussion is finally conventional, even if it triggered a debate (often acrimonious in style) that has endured for more than a half-century. Its conventionality—informed by a naïve scientific triumphalism and full of ill-judged predictions—remains a major liability for Snow, and for the afterlife of *The Two Cultures*.

And yet there is another dimension of Snow's argument—the contest of the art–nature relationship—that is less conventional. In order to pursue this other dimension, it is necessary that we think about crystallography a bit more rigorously. Snow's essay poses two great forces at work in history: the social forces dedicated to crystallizing established structures of privilege, and opposed to these, the sheer force of historical change itself. But what happens on a deeper level in Snow's work on the two cultures is something of a metamorphosis in which these two forces merge and become one. The working out of this metamorphosis extends beyond *The Two Cultures* and is continued in Snow's 1963 text *The Two Cultures: A second look*. In the four years between the Rede Lecture and *A Second Look* something fundamental has changed for Snow and I want to argue that the fundamental change is that by 1963 for Snow crystallography becomes the science of life itself with the startling realization that "crystals" and a particular kind of "crystallization" *constitute* life itself. In terms familiar to us from the previous chapter and its discussion of *The Winter's Tale* and seventeenth-century garden practices, DNA emerges for Snow as a perfect and, in fact, *defining* instance of the ways in which science leads us to understand that nature is indeed art.

III

In *A Second Look*, Snow begins by briefly (and, I would say, thoughtfully) commenting on the criticisms and debates that his original essay inaugurated, including an acknowledgment of the frequently hostile *ad hominem* nature of some of the more egregious attacks.[17] Once having run through what are in effect the preliminary matters, Snow turns his attention to what he calls a "passage where I showed bad judgement" when seeking to illustrate the severity of the problem of non-

---

[17] As he indicates in a note to this sentence, Snow is speaking here of Leavis (1962). Snow tells of having been asked in advance of publication for his consent, which he granted. Snow comments, "I had to assure them that I did not propose to take legal action. All this seemed to me distinctly odd. In any dispute acrimonious words are likely to fly about, but it is not common, at least in my experience, for them to come anywhere near the limit of defamation" (*Second Look*, 57).

communication between science and traditional culture. And it is here, I suggest, that we can detect the critical shift in Snow's argument about crystallization and change across the two essays. Although not in the least willing to retreat on his claims about the virtual non-existence of communication across the cultural divide—"if anything I understated the case" (*Second Look*, 71)—he is bothered by the choice of what he calls "my test question about scientific literacy, *What do you know of the Second Law of Thermodynamics?*" (*Second Look*, 71–2). While this is not exactly the form this question took in *The Two Cultures,* and although in this version the figure of Shakespeare has been—or *appears* to have been—excised altogether, Snow finds that by 1963 the field of thermodynamics as such is no longer an apt figure for the argument he wishes to make about "scientific literacy." For although it is, Snow says, "a good question," and though it is a law "of the greatest depth and generality: it has its own sombre beauty," the matter of the second law is (or has become) insufficiently communicable for Snow's purposes. But he is ready to provide a more promising candidate:

> I should now treat the matter differently, and I should put forward a branch of science which ought to be a requisite in the common culture, certainly for anyone now at school. This branch of science goes by the name of molecular biology. [*Second Look*, 72–3]

Snow then wastes no time explaining that "through a whole set of lucky chances," molecular biology is "ideally suited to fit into a new model of education" (*Second Look*, 73). The reason Snow believes so strongly in the potential for molecular biology to be, as it were, both the measure of and the means to scientific literacy is rooted in its status (I want to say) as a form of crystallography. Molecular biology, Snow argues, is "fairly self-contained":

> It begins with the analysis of crystal structure, itself a subject aesthetically beautiful and easily comprehended. It goes on to the application of these methods to molecules which have literally a vital part in our own existence—molecules of proteins, nucleic acids: molecules immensely large (by molecular standards) and which turn out to be of curious shapes, for nature, when interested in what we call life, appears to have a taste for the rococo. [*Second Look*, 73]

The event—which was scientific in its nature and cultural in its implications—that both triggers Snow's shift from thermodynamics to molecular biology and provides a kind of master discourse is the "leap of genius by which Crick and Watson snatched at the structure of DNA and so taught us the essential lesson about our genetic inheritance" (*Second Look*, 73). There is hardly a more significant moment in the entire two-cultures controversy. In this passage (and in the broader shift unfolding in Snow's understanding of science and culture more generally), Snow reinvents science as the collaborative work of nature and art, to the point, in fact, of their virtual indistinctness: when it considers life, nature, it turns out, has a taste for art.

The double-helix nature of DNA is the idea figure—or embodiment—of Snow's new understanding of nature itself, and the centrality of crystals (and crystallography) to this new dispensation. Rather than imagining something like inert matter (crystals, in our common understanding) as the material upon which science, or art, can work—processes that often go by the names of practical science and technology—molecular biology allows us to grasp a new fundamental law of nature, no less spectacular or beautiful or powerful than the laws of thermodynamics: that structure *itself* can be the "source" of life.

As a powerful example of scientific discoveries that are "closely connected to human flesh and bone," molecular biology and the double helix "are bound to touch both our hopes and our resignations":

> That is, ever since men began to think introspectively about themselves, they have made guesses, and sometimes had profound intuitions, about those parts of their own nature which seemed to be predestined. It is possible that within a generation some of these guesses will have been tested against exact knowledge. [*Second Look*, 75]

Molecular biology offers a new figuration for hope and Snow, always willing to offer a prediction, is willing to wager a guess: "No one can predict what such an intellectual revolution will mean: but I believe that one of the consequences will be to make us feel not less but more responsible towards our brother men" (*Second Look*, 75). And because molecular biology (unlike thermodynamics) "does not involve serious conceptual difficulties" and "needs very little mathematics to understand," it is more universally available. "What one needs most of all," Snow writes, "is a visual and three dimensional imagination." Molecular biology, he concludes, "is a study where painters and sculptors could be instantaneously at home" (*Second Look*, 73). I would like to mark this notion of the suitability—the *natural* suitability—of molecular biology to the talents and abilities of painters and sculptors for a more careful and sustained reading later in this chapter. What is "a visual and three dimensional imagination"? Why do painters and sculptors have a special claim on its proper possession and use? And, how does the emergence of molecular biology in Snow's two-cultures argument relate to the second law?

The final sections of *A Second Look* are dedicated to the unfolding of the consequences of Snow's new understanding of the two cultures. It becomes clear, for example, that the paired cultures at the very heart of the discussion are (as was suggested above) the rich and the poor. "Here, in fact, was what I intended to be the centre of the whole argument. Before I wrote the lecture I thought of calling it 'The Rich and the Poor', and I rather wish that I hadn't changed my mind" (*Second Look*, 79). Another clarification that emerges in *A Second Look*, I would argue, is the focus on history. Snow is quite consistent in arguing that the emergence of the industrial–scientific revolution was an absolute good; that the great social change produced by the revolution was both an unparalleled social good and is still unfolding, albeit unevenly, across the two global cultures of rich and poor; and that in light of these benefits provided by something like the historical march of science, protests against science are irrational because they are *unhistorical*:

> There is a mass of other evidence, from many kinds of provenance, all pointing in the same direction. In the light of it, no one should feel it seriously possible to talk about a pre-industrial Eden, from which our ancestors were, by the wicked machinations of applied science, brutally expelled. When and where was this Eden? Will someone who

hankers after the myth tell us where he believes it was located, not in terms of wishful
fancy, but in place and time, in historical and geographical fact? Then the social historians
can examine the case and there can be a respectable discussion. [*Second Look*, 83–4]

But the implication is clear: there cannot be a respectable discussion of this point.
Indeed, the logic of Snow's argument does not so much object to the idea of Eden
as such, but rather to its location in our mythic past, as seems to happen (he
suggests) in standard literary or traditional thinking about science. Snow's vision of
history is itself progressivist and utopian in nature and Eden awaits our greater acts
of selflessness and altruism on behalf of our "common humanity:"

> Millions of individual lives, in some lucky countries like our own, have, by one gigantic
> convulsion of applied science over the last hundred and fifty years, been granted some
> share of the primal things. Billions of individual lives, over the rest of the world, will be
> granted or will seize the same. This is the indication of time's arrow. [*Second Look*, 85]

It is critical to Snow's historical argument that the lives so vastly improved by the
industrial–scientific revolution are understood to be *individual* lives. At the same time,
however, these countless individual lives become great only when understood as
linked together through the "common humanity" Snow often mentions, or through
the progress of time figured, with great ambiguity (as I will argue presently) as "time's
arrow." It follows, then, that the collective (always understood to be composed of
particulars and individuals) should take as its founding political action the resistance of
what Snow calls "the individual will" that has long been the support of the crystallization so detrimental to the common good. Snow identifies an indeed disturbing passage
from D. H. Lawrence's *Studies in Classic American Literature*: Lawrence's celebration
of the famous flogging scene in Dana's *Two Years before the Mast*. For Lawrence, the
Captain's merciless flogging of Sam—which he sees as a figuration of the proper
master–servant relationship—is therapeutic and restorative. What happens in the
flogging of the sailor? "The living nerves respond," Lawrence writes:

> They start to vibrate. They brace up. The blood begins to go quicker. The nerves begin to
> recover their vividness. It is their tonic. The man Sam has a new clear day of intelligence,
> and a smarty back. The Captain has a new relief, a new ease in his authority, and a sore
> heart.
>   There is a new equilibrium, and a fresh start. The *physical* intelligence of a Sam is
> restored, the turgidity is relieved from the veins of the Captain.
>   It is a natural form of human coition, interchange.
>   It is good for Sam to be flogged. It is good, on this occasion, for the Captain to have
> Sam flogged. I say so. [Lawrence, qtd. in *Second Look*, 88]

Snow's rejection of Lawrence's vision is unequivocal: "This reflection," he writes,
"is the exact opposite of that which would occur to anyone who had never held, or
expected to hold, the right end of the whip—which means most of the poor of the
world, all the unprivileged, the teeming majority of our fellow men" (*Second Look*,
88–9). Snow then draws the proper—and, one could say, humanitarian—response:
"as soon as the poor began to escape from their helplessness, the assertion of the
individual will was the first thing they refused to take" (*Second Look*, 89).

Snow concludes *A Second Look* with a humanitarian and educational appeal. And while he admits that "we are not going to turn out men and women who understand as much of our world as Piero della Francesca did of his, or Pascal, or Goethe," we can nevertheless educate "a large proportion of our better minds so that they are not ignorant of imaginative experience, both in the arts and in science, nor ignorant either of the endowments of applied science, of the remediable suffering of most of their fellow humans, and of the responsibilities which, once they are seen, cannot be denied" (*Second Look*, 100).

I would like to return to Snow's use of the phrase "time's arrow," which he deploys in order to argue for something like the historical inevitability *in the future* for a number of social and political eventualities: the more equitable global distribution of both science and wealth, for example, or the improvement of the general quality of life of the world's poor. For Snow, both of these unequivocal social goods will take place—both because they are now possible and because educated Western "man" will come to see the rightness of these two global realignments and in the spirit of brotherhood and common humanity work for their realization.[18] But Snow (like the rest of us) has the notion of "time's arrow" not from an underlying Hegelian notion of the gradual but inexorable march of spirit toward absolute realization, but rather from the second law of thermodynamics. And the phrase itself—time's arrow—was introduced by the Cambridge mathematician and physicist, A. S. Eddington (whose name in fact appears in *The Two Cultures* in a list of the high-caliber mathematical minds made to endure the old Mathematical Tripos at Cambridge). Eddington is credited with (among other firsts) the introduction of the phrase "time's arrow." And as the following brief passage makes quite clear, for Eddington, the direction of the arrow indicates greater or lesser degrees of disorder and can therefore indicate the proper "location" of both past and future:

Let us draw an arrow arbitrarily. If as we follow the arrow we find more and more of the random element in the state of the world, then the arrow is pointing towards the future; if the random element decreases the arrow points towards the past . . . I shall use the phrase "time's arrow" to express this one-way property of time which has no analogue in space.[19]

---

[18] Snow offers a rough estimate of the sheer scale of the problem—and the scale of the necessary solution—in human terms: "Imagine," he writes, "that the U.S. government and ours had agreed to help the Indians to carry out a major industrialisation, similar in scale to the Chinese. Imagine that the capital could be found. It would then require something like ten thousand to twenty thousand engineers from the U.S. and here to help get the thing going" (*Two Cultures*, 47). In addition to the challenge of finding that many engineers to throw themselves into the task, Snow cautions that they "would need to be trained not only in scientific but in human terms. They could not do their job if they did not shrug off every trace of paternalism." Such new men, he argues, would avoid the pitfalls of European paternalism, "do an honest technical job, and get out." As it happens, "this is an attitude which comes easily to scientists" because they, among all of us:

are freer than most people from racial feeling; their own culture is in its human relations a democratic one. In their own internal climate, the breeze of the equality of man hits you in the face, sometimes rather roughly, just as it does in Norway. [*Two Cultures*, 47–8]

[19] Eddington (1928, rpt. 1958: 69). Eddington's book is, he writes at the outset, "substantially the course of Gifford Lectures" that he delivered at the University of Edinburgh in 1927 (p. vi).

Snow certainly knew Eddington's work and the phrase for which he is still famous. But is Snow's use of "time's arrow" an accurate or apt one? I suspect that it is neither, but this will be difficult to demonstrate because—on the most basic level— Snow's use of time's arrow (as a figure for the second law) is complicated by the fact that the figure itself is meant to describe a physical phenomenon (for simplicity's sake we can call this phenomenon "entropy") that by its very nature resists easy application to the social (or, the artifactual) world. There are a number of possibilities, however, embedded within Snow's use of time's arrow. First, he may mean to argue that an equitable world—greater and more even distribution of wealth and privilege—will necessarily result from the passage of time because wealth, say, or political power, will eventually be dispersed more broadly (or, in Eddington's phrasing, more *randomly*) than it has historically by virtue of remaining in a state of high concentration within an extremely narrow domain of the world's nations and their populations. As such, the historical tendency of wealth to remain condensed or concentrated (let us say) and not broadly dispersed or randomized is a thoroughly unnatural state which requires the expenditure of vast amounts of energy to sustain against the natural entropic pull. These vast amounts of energies have taken the form of political and economic oppression, as well as those forces of social crystallization which Snow finds so objectionable.

A second possibility for the meaning of Snow's use of time's arrow is based upon the same thermodynamic understanding of other social (and even individual) characteristics, such as morality and benevolence, selflessness and charity. On Snow's account, all of these desirable social characteristics will also become broadly or more randomly dispersed across national and, then, global cultures. They will no longer remain confined (through the exertion of certain energies) to particular and isolated groups. Instead, the social world will experience an inevitable spread of morality or selflessness, even as heat (unless constrained by a continuous application of energy) will disperse from areas of greater to areas of lesser concentration.

A third possible meaning of Snow's arrow would argue that the forward motion of time—since Snow never could contest the inexorable progression of time in a forward direction guaranteed by the laws of thermodynamics—will bring with it the gradual dispersion of precisely those energies (oppression, as an example, or armies) required to contain wealth and power in their historical locations. One nightmare version of this scenario conjures a state of fundamental anarchy and violence not unlike the one imagined by Hobbes (and echoed in *The Two Cultures*) prior to the social contract, a state of an absolutely dispersed power that guarantees life to be nasty, brutish, and short (*Two Cultures*, 42). But this is not the future imagined hopefully in Snow's two cultures; that future is one of dispersed (and random) benevolence and generosity.

But even as the debate obtains concerning the number of cultures—one, two, three, or "two thousand and two"?—so one could multiply the possible optimistic meanings potentially deployed by Snow's arrow: the spread of benevolence, wealth, selflessness, "common humanity," power, privilege, access to science and technology, and so forth.[20] The problem, however, is that there is, strictly speaking, nothing in Snow's model to distinguish any of these from another, nor any of these from something like their opposites. Indeed, the movement of time itself imagined in Snow's argument may just as likely be said to lead inexorably to the gradual dispersion of all of those negative and destructive individual and social characteristics Snow's *Two Cultures* means to resist: the spread of human suffering, the dispersion of poverty across the globe, the increasing inability of the two cultures to speak to each other (whether they are imagined as science and tradition, or rich and poor). The great liability of time's arrow when applied to the social world (and not to clouds of atoms or thermodynamic systems) is that it tends to predict chaos—not the image Snow needs his analysis to envision or announce.

These multiple possibilities and, more to the point, our inability to decide among them based upon the paradigm provided by thermodynamics, arise (as I have suggested) from the application of physical laws to social-artifactual entities in a non-tropological fashion. In other words, while the second law of thermodynamics is indeed immensely powerful and applicable to the material world as we now know it, when it comes to its application to the social (that is, the constructed or the artifactual) world in which we also live, the second law is immaterial at best, and potentially seriously prejudicial or harmful.

But, as I suggested above, Snow would have known all of this. Indeed, he was quite clearly aware of the intellectual dangers that attend the too-easy assumption that the physical world is modeled on our own immediate experiences of it. One telling instance from *The Two Cultures* where this is made evident is in Snow's invocation of the non-conservation of parity discovered in 1957, which Snow calls "one of the most astonishing discoveries in the whole history of science was brought off. I don't mean the sputnik—that was admirable for quite different reasons, as a feat of organisation and a triumphant use of existing knowledge. No, I mean the discovery at Columbia by Yang and Lee." He continues, in something like genuine wonderment:

It is a piece of work of the greatest beauty and originality, but the result is so startling that one forgets how beautiful the thinking is. It makes us think again about some of

---

[20] In *A Second Look*, Snow addresses criticism of his identification of *two* cultures. On the one hand, he points to the various arguments for a different number of cultures (there are critics, he writes, who complain that "there aren't two cultures, there are a hundred and two, or two thousand and two, or any number you like to name. In a sense this is true: but it is also meaningless" [*Second Look*, 66]). On the other hand (and more seriously), Snow admits having been "slow to observe the development of what, in the terms of our formulae, is becoming something like a third culture" (*Second Look*, 70)—the emergence of what we call the social sciences:

I think also that writing as an Englishman made me insensitive to something which may, with a few years, propel the argument in another direction or which conceivably may already have started to do just that. I have been increasingly impressed by a body of intellectual opinion, forming itself, without organisation, without any kind of lead or conscious direction, under the surface of this debate . . . This body of opinion seems to come from intellectual persons in a variety of fields—social history, sociology, demography, political science, economics, government (in the American academic sense), psychology, medicine, and social arts such as architecture. [*Second Look*, 69–70]

the fundamentals of the physical world. Intuition, common sense—they are neatly stood on their heads. [*Two Cultures*, 15]

The non-conservation of parity—the demonstration that nature, on the subatomic level (and, more particularly, within the dynamic of the weak force and weak interactions), indeed violates our common-sense (or, natural) assumptions about a right–left symmetry—serves in Snow's account to remind us that nature does not function in accord with our assumptions about it. But this reminder seems to be at odds with the deployment of time's arrow to describe the future state of the social world if by that use the social world is imagined as organized by physical laws. In other words, Snow's use of time's arrow to describe or predict social states offers the curious instance of the application of the second law of thermodynamics violating the second law of thermodynamics.

But does social crystallography represent a way out of this problem? Can molecular biology solve this apparent collapse of the second law of thermodynamics when cast within the social dimension? We have already seen how Snow effectively displaces the paradigm of thermodynamics with that of molecular biology as both the ideal measure of scientific literacy and as the intellectual model for both cultures. I argued above that the privileging of molecular biology arises from Snow's intuition, born of his new awareness of the discovery (by Crick, Watson, and—as we know now—Franklin) through x-ray crystallography of the structure of DNA, that structure itself is the machine that generates life. And even more than this, molecular biology—and DNA, with its double-helix structure, which is its sign—emerges as that structure, which is natural in nature (so to say), that can be said to have governed history (Snow refers to the discovery of the double helix as that "leap of genius by which Crick and Watson . . . taught us the essential lesson about our genetic inheritance" [*Second Look*, 73]). At the same time, this structure can also be said to govern the future, as the double helix carries our genetic information into a future that it has already (genetically) determined.[21]

This means that in Snow's argument, molecular biology not only takes the place of the second law of thermodynamics as the measure of scientific literacy and the method for the communicability of science; it also—and more importantly—serves *to replace* time's arrow altogether as that mechanism, harvested from the culture of science, that helps explain the nature of human existence and experience and that, at the same time, marks the distinction between the two cultures with an seemingly absolute precision and certainty. It is perhaps for this reason that Snow believes molecular biology will be especially attractive to those among us with a particularly highly-developed "visual and three dimensional imagination." It is the world produced by molecular biology in which "painters and sculptors could be instantaneously

---

[21] For a powerful and compelling discussion of a number of these issues, see Turner (2007). Turner's book—which, among other goals, wishes "to advance a simple but counter-intuitive argument: that we should regard genetic engineering and biotechnology not simply as a new application of scientific knowledge but rather as a new mode of poetics, and that Shakespeare's own work provides a model for just such an approach" (p. 7)—will be required reading for students of early modern theater (Shakespeare in particular) and emergent science.

at home" (*Second Look*, 73). But since scientists, those people who have brought molecular biology to us in the first place, presumably already have and exercise this visual and three dimensional imagination, Snow's welcoming announcement of this new world is an elaborate way of issuing an invitation to the artist to join. The invocation of the artist, that is, stands as one powerfully clear instance in Snow's entire two-cultures discussion in which the two cultures are understood fundamentally to be the "culture" of nature and the "culture" of art. And as the figures of the painters and sculptors suggest, the artists in Snow's world are always belated, and their talents are always in the service of reflecting the natural world.

What is troubling about Snow's linking of Shakespeare and thermodynamics—and this gesture is itself a sign of Snow's understanding of the relationship of art to nature—is the recognition of a certain tension between a natural world that is itself understood to be a work of art and a work of art that is understood to be as natural as the laws of thermodynamics. Snow's molecular biology discovers a "rococo" world at fundamental levels, even as his crystallography displaces the second law of thermodynamics precisely because only art can be imagined to work against the directionality of time's arrow and thereby amend nature.

8

Conclusion

Being Archaic

We are archaic in three-fourths of our actions. Few people and even fewer thoughts are completely congruent with the date of their times.

Michel Serres

It has been the central concern of this book to interrogate the traditional, though deeply problematical, separation of the disciplines and practices of science from the rest of culture. My primary texts are early modern ones, both literary and scientific in nature—in the conventional (that is to say, inadequate) senses of those terms. For it has been of great importance throughout this study to demonstrate the ways in which these categories are themselves products of a complicated and highly successful series of disciplinary strategies and maneuvers dedicated to establishing and carefully policing the division between science and culture, or between science and literature, or between nature and art. Thus, if the early modern writers whose works have been under discussion here—Francis Bacon, William Shakespeare, Galileo Galilei, John Donne, John Evelyn (as well as Michel de Montaigne, Johannes Kepler, Tycho Brahe, George Wither, and Robert Greene, among others)—can be said to have inhabited a moment prior to the establishment of science as a separate sphere, then it is even more apposite within the context of this project to say, too, that they lived and wrote in a moment in which there was no strict division between the "two cultures," a moment (as a number of critics have recently argued) in which literary and natural philosophical endeavors were in significant ways functionally inseparable. This, at least, has been the contention I have offered in the preceding discussions of these writers—a term that I prefer to "dramatist," or "poet" or, indeed, "scientist" precisely because it marks a certain and useful resistance to rigid generic taxonomical imperatives that themselves bear the marks of disciplinarity.

At the same time as I have read early modern writers, I have also been keen to engage with a set of writers who come to us from the middle decades of the twentieth century, including writers as diverse as T. S. Eliot, Walter Benjamin, Marjorie Hope Nicolson, C. P. Snow, William Empson, and (a bit later) Clifford Geertz. Perhaps, then, a word of explanation (after the fact) may be in order. On the one hand, there are obvious reasons why a book that addresses early modern literature and science would be interested in the work of Nicolson or Empson, for example: both critics were themselves deeply—and productively—engaged in their own attempts to understand these two great discourses and the relationship (as they saw it) between them. At the same time, if our subject is the nature of experience at the opening of the moment of what many call the modern era, then turning to Eliot or, especially, to Benjamin, both of whom are theoreticians of modernity, makes a certain obvious sense.

But there are, on the other hand, a number of less obvious though nevertheless important reasons that this set of significant secondary texts should emerge in this study. Most importantly is that from a particular perspective—a perspective that is made available in part through the practices of science studies—the early modern texts and the "modern" texts seem to solicit each other because they are *proximate* to one another. My point here is twofold. First, early modern and modern texts are proximate if we consider them linked in an historically significant fashion: as framing moments, for instance, in which the thing thus framed may well be the virtually complete separation of science from culture that is eventually achieved between the seventeenth century and the nuclear age—at the point, that is, of an apparent dominance of science. Secondly, the early modern and the modern can be proximate moments if we take up the alternative understanding of time, and, by implication, of history, held out in Michel Serres's theorizations of science and, more particularly, the experience of experiencing time. Arguing that once one enters into a topological understanding of time and the experience of time, one comes immediately to "realize how much all of what we've said about time up till now abusively simplifies things," Serres offers something of a parable in order to illustrate this alternative conception of time figured topologically:

> [T]his time can be schematized by a kind of crumpling, a multiple, foldable diversity.... We are always simultaneously making gestures that are archaic, modern, and futuristic. Earlier I took the example of a car, which can be dated from several eras; every historical era is likewise multitemporal, simultaneously drawing from the obsolete, the contemporary, and the futuristic. An object, a circumstance, is thus polychronic, multitemporal, and reveals a time that is gathered together with multiple pleats.
>
> If you take a handkerchief and spread it out in order to iron it, you can see in it certain fixed distances and proximities. If you sketch a circle in one area, you can mark out nearby points and measure far-off distances. Then take the same handkerchief and crumple it, by putting it in your pocket. Two distant points suddenly are close, even superimposed. If, further, you tear it in certain places, two points that were close can become very distant. This science of nearness and rifts is called topology, while the science of stable and well-defined distances is called metrical geometry.[1]

Noting (after Bergson) that "Classical time is related to geometry, having nothing to do with space," Serres offers a definition of the nature of what he calls "crumpled time"—a concept that is not only intended as a heuristic tool, but that also is meant to describe the (temporal) world as we actually encounter it: "As we experience time—as much in our inner senses as externally in nature, as much as *le temps* of history as *le temps* of weather—it resembles this crumpled version much more than

---

[1] Serres and Latour (1995: 59–60).

the flat, overly simplified one. Admittedly, we need the latter for measurements, but why extrapolate from it a general theory of time? *People usually confuse time and the measurement of time*, which is a metrical reading on a straight line" (pp. 60–1, emphasis in the original).

Serres's crumpled time, as these passages make clear, revises—by supplementing—our traditional understanding of time and how exactly we experience it. While this theorization may seem to be archly modern—or postmodern, perhaps—to some readers, it is more properly understood as *archaic*, Serres's term meant to signify the polychronic nature of objects and things in the world—cars (for example) or theories of time. "Archaisms can always be found among us," Serres writes, "while Lucretius, in some instances, is right on top of things, as they say" (p. 61).

This idea of topological time (since it is, after all, archaic) is familiar to us from a number of its early modern analogs, one of which is the practice of chorography: the careful description of place as it exists in multiple points in history on display in such works as William Camden's *Britannia*, John Speed's *Theatre of the Empire of Great Britain*, and Michael Drayton's *Poly-Olbion*.[2] Or we see it in theologically inflected meditations on Christian salvation, such as John Donne's great poem "Hymn to God my God, in my Sickness":

> We think that Paradise and Calvary,
> Christ's Cross, and Adam's tree, stood in one place;
> Look Lord, and find both Adams met in me;
> As the first Adam's sweat surrounds my face,
> May the last Adam's blood my soul embrace.[3]

One scholar of early modern literature and culture who has written powerfully on topological time is Jonathan Gill Harris, whose recent book, *Untimely Matter in the Time of Shakespeare*, shares with the present study an abiding concern with the notions of the polychronic and the multitemporal, as theorized not only by Serres and Latour, but also by a range of early modern writers. Although Harris is equally interested in what might be considered a new form or version of object studies, his understanding of materiality—or matter—is itself the product of a fruitful cross-fertilization with re-theorizations of temporality. "A strikingly wide array of six-teenth- and seventeenth-century forms of English and literary activity—devotional lyric verse, urban chorography, vitalist philosophy, and, most insistently, Shake-speare's own drama—expound or enact theories of the polychronic nature of matter."[4] In one of his most compelling chapters, Harris considers the famous

---

[2] For a discussion of early modern chorography, see Marchitello (1997, esp. Ch. 3); see also Helgerson (1992).

[3] Donne, "Hymn to God my God, in my Sickness" in Donne (1990: 332).

[4] Harris (2009: 4). Harris continues,

These theories range from the explicit to the implicit, the philosophical to the practical, from the religious to the secular. But each understands matter to collate diverse moments in time. In the process, these theories do more than just recognize the polychronicity of the object; they also insist on its multitemporal properties—that is, its materialization of diverse relations among past, present, and future. [p. 4]

"double time" problem said (at least coventionally) to afflict *Othello*. In his discussion, called "Crumpled handkerchiefs," Harris argues that the play is not "afflicted by a temporal anomaly" but rather "refuses linear temporality" (p. 169). The alternate temporality of the play, Harris argues, is "materialized in *Othello*'s most untimely stage property: the handkerchief." Citing the multiple exchanges of the handkerchief (from Othello to Desdemona, from Desdemona to Emilia, from Emilia to Iago, from Iago to Cassio, from Cassio to Bianca and back again, and so forth until it returns once again to Othello at the end), as well as the handkerchief's multiple genealogies ("antique Egyptian token and disposable European trifle, old pagan fetish and New Testament instrument of healing . . . "), Harris identifies the handkerchief's—and the play's—temporal theorization:

> By quilting together old and new, pre- and post-, past and present—just as the play (at least in Iago's imagination) knots "old black ram" and "young ewe" into the infamous "beast with two back"—the handkerchief hints at how *Othello* refuses temporal as much as racial purity. Rather, the play, like the handkerchief, trades in an impure, preposterous temporality that we might call crumpled time. [p. 170]

We encounter still another version of topological time in the famous gravedigger scene in *Hamlet*. Indeed, *Hamlet*'s gravedigger provides a perfect—and perfectly comic—apotheosis of that play's obsession with de-cultivation: the perhaps tragic movement from garden to grave where presides not growth but death, not order but disorder, not composition but decomposition. From Old Hamlet's "orchard" (the site of his customary repose) to Hamlet's "unweeded garden" (a site of utter neglect) to the gravedigger's "house" (that product of his intense labors that "lasts till doomsday"), the play tends toward de-cultivation.[5] Even as he stands in his grave, the gravedigger recalls Adam and harkens back to the primal garden that seems here to devolve into the grave: "There is no ancient gentlemen but gardeners, ditchers, and grave-makers—they hold up Adam's profession." When pressed— "Was he a gentleman?"—the gravedigger replies, "A was the first that ever bore arms." And when pressed again—"Why, he had none"—he concludes, "What, art a heathen? How dost thou understand the Scripture? The Scripture says Adam digged. Could he dig without arms?" (5.1.29–37).

One powerful effect of this scene is its renegotiation of time as strictly linear (metrical): it is effectively gentleman Adam, and not just the gravedigger, who digs the grave and scatters the bones to make way for a new tenant. History collapses in this allegory and Adam's garden decomposes into the open grave: Adam's new grim harvest.[6] "Your worm is your only emperor for diet," Hamlet has already declared, "we fat all creatures else to fat us, and we fat ourselves for

[5] Shakespeare (1982: 1.5.35, 1.2.135, 5.1.59). Even these terms/ideas themselves, however, tend toward decomposition: Old Hamlet's "orchard" proves less than impregnable and his "secure" hour is precisely the moment of his death. Similarly, the gravedigger's notion that the houses he builds last till doomsday is given the lie by none other than the gravedigger himself who breaks into such a house when he digs what will become Ophelia's clearly temporary grave.

[6] When asked where Polonius is, Hamlet replies quite knowingly that he is at supper: "Not where he eats, but where a is eaten. A certain convocation of politic worms are e'en at him" (4.3.19–20).

maggots. Your king and your lean beggar is but variable service—two dishes, but to one table. That's the end" (4.3.21–5). But even as this grave gapes and crumples time, it also serves as a prompt to Hamlet and incites his further inquiry: it is, indeed, the grave beside which Hamlet will soon stand and into which he will peer with such serious—and materialist—curiosity: "How long will a man lie i'th'earth ere he rot?" (5.1.158).

On the level of methodology, we encounter an analog of Serres's crumpled time in the work of Bruno Latour, one of the leading figures in science studies today. [7] In his book *We Have Never Been Modern*, Latour offers a sustained—and highly critical—analysis of the emergence (in the wake of Kant) of what he calls the "modern Constitution" that is characterized by a double motion: toward the establishment of the very notion of the wholly separate spheres of Nature and Culture, on the one hand, and the simultaneous cancellation of this absolute polarity, on the other. "What an enormous advantage," Latour writes, "to be able to reverse the principles without even the appearance of a contradiction!" (p. 37).[8]

For Latour, who labels his project as work toward an "anthropology of the modern world" (p. 15), it is modernity itself, established by this Constitution, that has made it impossible for us to see clearly the "confused mixture" of technology and society identified by Serres—or, in Latour's particular vocabulary, those "hybrid" entities that exist neither wholly on the Nature pole nor wholly on the Society pole posited by the modern Constitution, but rather somewhere both between them and off the line (spectrum) that connects them.[9] Citing the dispersed or hybrid nature of the (social and scientific) problems of the spread of the AIDS virus, for example, or the discovery of the growing hole in the ozone layer, Latour argues that the modern Constitution and its disciplinary practices work rigorously (if silently and behind the scenes) to simplify its complexities and to submerge the interconnectedness of Nature ("things") and Society ("power and human politics"):

> Press the most innocent aerosol button and you'll be heading for the Antarctic, and from there to the University of California at Irvine, the mountain ranges of Lyon, the chemistry of inert gases, and then maybe to the United Nations, but this fragile thread will be broken into as many segments as there are pure disciplines. By all means, they seem to say, let us not mix up knowledge, interest, justice and power. Let us not mix up

---

[7] Latour (1993); see also (1987), (1999), (2004).

[8] Latour continues:

> In spite of its transcendence, Nature remains mobilizable, humanizable, socializable. Every day, laboratories, collections, centres of calculation and of profit, research bureaus and scientific institutions blend it with the multiple destinies of social groups. Conversely, even though we construct Society through and through, it lasts, surpasses us, it dominates us, it has its own laws, it is as transcendent as Nature. For every day, laboratories, collections, centres of calculation and of profit, research bureaus and scientific institutions stake out the limits to the freedom of social groups, and transform human relations into durable objects that no one has made. The critical power of the moderns lies in this double language: they can mobilize Nature at the heart of social relationships, even as they leave Nature infinitely remote from human beings; they are free to make and unmake their society, even as they render its laws ineluctable, necessary and absolute. [p. 37]

[9] See Latour (1993, esp. Ch. 3) and his discussion of what he calls (following Michel Serres) "quasi-objects" and "quasi-subjects."

heaven and earth, the global stage and the local scene, the human and the nonhuman. "But these imbroglios do the mixing," you'll say, "they weave our world together!" "Act as if they didn't exist," the analysts reply. They have cut the Gordian knot with a well-honed sword. The shaft is broken: on the left, they have put knowledge of things; on the right, power and human politics. [pp. 2–3]

For both Serres and Latour, the task before science studies is to understand these networks, even in the face of the historical and prevailing desire of modernity radically to dismantle them. But if the problems analyzed by Serres and Latour are linked to modernity (and this is especially true for Latour), and if both advocate discussion of networks before they are dismantled by modern disciplinarity and scientism, then one very promising avenue of study to pursue is the investigation of science and society (to use these general terms) literally *prior to* the advent of the Latour's modern Constitution.

It has been the aim of *The Machine in the Text* to participate in just this project: to try to think about science before the modern—and, more particularly, to think about the relationship between early modern science and literary culture without accepting the historical argument of the two cultures. As I have indicated at a number of moments in this book, the project of rethinking early modern science and culture is an urgent concern for many critics today—in new readings of art as a form of knowledge, of early modern technological culture, of machines, of botanical and horticultural practices as creative and imaginative discourses, of art as nature and of nature as art.

In many ways, these new studies surprise us when we see, say, Shakespeare on the same page with the double helix, or perhaps when we see Francis Bacon in dialog (as it were) with both Christmas festivities in 1594 and with Geertz's version of anthropology and x-ray diffraction patterns. But perhaps the real surprise here is that this criticism is, in the end, perfectly familiar to us—at least in kind. For the work offered here (and surveyed here) is another form, though not so familiar, of historicism. It bears a striking family resemblance to the new historicism, and to the kind of new critical idiom tentatively proposed by Henry S. Turner's invocation of a "new new historicism" in his analysis of the concept of "life" in modern genetic science and *A Midsummer Night's Dream*.[10] For after all, perhaps the great lesson of crumpled time is to know that being archaic is another way of being in history.

---

[10] Turner (2009: 199).

References

PRIMARY

(Place of Publication is London unless otherwise noted.)

Anon. (1688), *Gesta Grayorum*, ed. W. W. Greg (Oxford: Malone Society Reprints/Oxford University Press).

Anon. (1688), *Gesta Grayorum: Or, The History of the High and Mighty Prince Henry Prince of Purpoole, Anno Domini 1594*, ed. Desmond Bland (Liverpool: Liverpool University Press, 1968).

Anon. (1595), *Gesta Romanorum: A Record of Auncient Histories Newly Perused by Richard Robinson* (Delmar, New York: Scholars' Facsimiles & Reprints, 1973).

Bacon, Francis (2000), *The New Organon*, eds. L. Jardine and M. Silverthorne (Cambridge: Cambridge University Press).

—— (1996), *The Oxford Authors: Francis Bacon: A critical edition of the major works*, ed. B. Vickers (Oxford: Oxford University Press).

Birch, Thomas (1756–7), *The History of the Royal Society of London for the Improving of Natural Knowledge*.

Brahe, Tycho (1598), *Astronomiae Instauratae Mechanica* (Wandensburgh).

—— (1946), *Tycho Brahe's Description of his Instruments and Scientific Work*, trans. and ed. Hans Raeder, Elis Strömgren, and Bengt Strömgren (København: I Kommission Hos Enjar Munksgaard).

Browne, Sir Thomas (1977), *Sir Thomas Browne: The Major Works*, ed. C. A. Patrides, (Hammondsworth: Penguin).

Buonamici, Francesco (1591), *De Motu* (Florence).

Cavendish, Margaret (2001), *Observations upon Experimental Philosophy*, ed. Eileen O'Neill (Cambridge University Press).

Charron, Pierre (1607), *Of Wisdom Three Books*, trans. Samson Lennard (Amsterdam and New York: Da Capo Press: Theatrum Orbis Terrarum, 1971).

Crooke, Helkiah (1615), *Microcosmographia*.

Descartes, René (1985), *The Philosophical Writings of Descartes*, 2 vols., ed. and trans. John Cottingham, Robert Stoothoff, and Dugald Murdoch (Cambridge: Cambridge University Press).

Donne, John (1987), *Devotions upon Emergent Occasions*, ed. Anthony Raspa (Oxford: Oxford University Press).

—— (1969), *Ignatius His Conclave*, ed. T. S. Healy (Oxford: The Clarendon Press).

—— (1651), *Letters to Severall Persons of Honour*, ed. M. Thomas Hester (Delmar, New York: Scholars' Facsimiles and Reprints, 1977).

—— (1990), *The Oxford Authors: John Donne*, ed. J. Carey (Oxford: Oxford University Press).

—— (1993), *Pseudo-Martyr*, ed. Anthony Raspa (Montreal: McGill-Queens University Press).

Evelyn, John (1706), *Acetaria*.

—— (2001), *Elysium Britannicum; Or, The Royal Gardens*, ed. John Ingram (Philadelphia: University of Pennsylvania Press).

—— (1706), *Kalendarium Hortense.*

—— (1664), *Sylva.*

—— (1995), *The Writings of John Evelyn*, ed. Guy de la Bédoyère (Woodgridge: The Boydell Press).

Galilei, Galileo (1967), *Dialogue Concerning the Two Chief World Systems*, 2nd edn, ed. Stillman Drake (Berkeley: University of California Press).

—— (1960), *Discourse on Bodies in Water*, ed. Stillman Drake, trans. Thomas Salusbury (1663) (Urbana: University of Illinois Press).

—— (1957), *Discoveries and Opinions of Galileo*, ed. Stillman Drake (New York: Anchor Books).

—— (1890–1909), *Le Opere di Galileo Galilei*, ed. Antonio Favaro, 20 vols. (Florence: Barbera).

—— (1989), *Sidereus Nuncius; Or, The Sidereal Messenger*, ed. and trans. Albert van Helden (Chicago: University of Chicago Press).

Green, Robert (1996), "Pandosto: The triumph of time" in Stephen Orgel (ed.), *The Winter's Tale*, (Oxford: Oxford University Press), 234–74.

Harvey, William (1653), *Anatomical Exercitations Concerning the Generation of Living Creatures.*

Hobbes, Thomas (1969), *Leviathan*, ed. Richard Tuck (Cambridge: Cambridge University Press).

Horky, Martin (1610), *Brevissima Peregrinatio Contra Nuncium Sidereum . . .* (Mantua).

James I (1994), *The Political Writings*, ed. Johann P. Sommerville (Cambridge: Cambridge University Press).

Jonson, Ben (1969), *Ben Jonson: The complete masques*, ed. Stephen Orgel (New Haven: Yale University Press).

Kepler, Johannes (1967), *Kepler's Somnium, Or, The Dream, Or, Posthumous Work on Astonomy*, ed. Edward Rosen (Madison: University of Wisconsin Press).

—— (1965), *Kepler's Conversation with Galileo's Sidereal Messenger*, ed. Edward Rosen (New York and London: Johnson Reprint Corporation).

Laurentius, Andreas (1599), *A Discourse of the Preservation of the Sight: of Melancholike Diseases; of Rheumes, and of Old Age*, trans. Richard Surphlet.

Maggi, Gerollamo (Hieronymus Magius) (1608), *Tintinnabulis.*

Marvell, Andrew (1990), *Andrew Marvel*, eds. Frank Kermode and Keith Walker (Oxford: Oxford University Press).

Montaigne, Michel de (2003), *Apology for Raymond Sebond*, trans. Roger Ariew and Marjorie Grene (Indianapolis and Cambridge: Hackett Publishing Company).

Pliny (1601), *The Historie of the World. Commonly called, The Naturall Historie of C. Plinius Secundus*, trans. Philemon Holland.

—— (1968), *Natural History, in Ten Volumes*, trans. H. Rackham (Cambridge: Harvard University Press and London: William Heinemann Ltd).

Pope, Alexander (ed.) (1723–5), *The Works of Mr. William Shakespear.*

Robinson, Richard (1595), *A Record of Auncient Hisories, intituled in Latin: Gesta Romanorum.*

Scheiner, Christoph (1630), *Rosa Ursina sive Sol ex admirando facularum & macularum* (Bracciano).

—— (1612), *Tres Epistolae de Maculis Solaribus Sciptae ad Marcum Welsrum.*

Shakespeare, William (1982), *Hamlet*, ed. Harold Jenkins (London and New York: Methuen).

—— (1954), *The Tempest*, ed. Frank Kermode (London: Methuen).

Shakespeare, William (1996), *The Winter's Tale*, ed. Stephen Orgel (Oxford: Oxford University Press).

Sidney, Philip (1989), *Sir Philip Sidney*, ed. Katherine Duncan Jones (Oxford: Oxford University Press).

Vasari, Giorgio (1991), *The Lives of the Artists*, trans. Julia Conaway Bondanella and Peter Bondanella (Oxford and New York: Oxford University Press).

Walton, Izaak (1950), *The Lives of John Donne, Sir Henry Wooten, Richard Hooker, George Herbert and Robert Sanderson*, ed. George Saintsbury (London: Oxford University Press).

Waterhouse, Edward (1663), *Fortescutus Illustratus, Or, A Commentary on that Nervous Treatis De Laudibus Legum Anglie, written by Sir John Fortescue, Knight . . .*

Wither, George (1635), *A Collection of Emblemes, Ancient and Moderne*.

—— (1975), *A Collection of Emblemes, Ancient and Moderne*, ed. Rosemary Freeman (Columbia: University of South Carolina Press).

Wright, Thomas (1604), *The Passions of the Minde in Generall*.

SECONDARY

Adelman, Janet (1992), *Suffocating Mothers: Fantasies of maternal origin in Shakespeare's plays, "Hamlet" to "The Tempest"* (New York: Routledge).

Albanese, Denise (1999), *New Science, New World* (Durham: Duke University Press).

Bald, R. C. (1970), *John Donne: A life* (New York and Oxford: Oxford University Press).

Bann, Stephen (1981), "A description of Stonypath", *Journal of Garden History*, 1/2: 113–44.

Barthes, Roland (1972), "The structuralist activity" in *Critical Essays*, ed. and trans. Richard Howard (Evanston: Northwestern University Press), 213–20.

Belsey, Catherine (1995), "Love as trompe-l'oeil: Taxonomies of desire in *Venus and Adonis*", *Shakespeare Quarterly*, 46: 257–76.

Benjamin, Walter (1969), "The storyteller" in Hannah Arendt (ed.) *Illuminations*, trans. Harry Zohn, (New York: Schocken Books), 83–110.

Biagioli, Mario (ed.) (1999), *The Science Studies Reader* (New York: Routledge).

—— (2006), *Galileo's Instruments of Credit: Telescopes, images, secrecy* (Chicago: University of Chicago Press).

—— and Peter Galison (eds.) (2003), *Scientific Authorship: Credit and intellectual property in science* (New York: Routledge).

Blair, Tony (2006), "Britain's bright path to the future: Lit by the brilliant light of science", ⟨http://www.number10.gov.uk/Page10342⟩, unpaginated.

Bloor, David (1976), *Knowledge and Social Imagery* (Chicago: University of Chicago Press).

Bono, James (1995), *The Word of God and the Languages of Man: Interpreting nature in early modern science and medicine* (Madison: University of Wisconsin Press).

Booth, Sara Elizabeth and Albert van Helden (2001), "The Virgin and the telescope: The moons of Cigoli and Galileo" in Jürgen Renn (ed.), *Galileo in Context* (Cambridge: Cambridge University Press), 193–216.

Brown, Bill (2004), "Thing theory" in *Things* (Chicago: University of Chicago Press), 1–21.

Bulletin of the Atomic Scientists of Chicago: A Journal of Science and Public Affairs (1945), 1/1.

Bushnell, Rebecca (2003), *Green Desire: Imagining early modern English gardens* (Ithaca: Cornell University Press).

Butterfield, Herbert (1949), *The Origins of Modern Science, 1300-1800* (London: G. Bell).

Cain, Tom (2006), "Donne's political world" in Achsah Guibbory (ed.), *The Cambridge Companion to John Donne* (Cambridge: Cambridge University Press), 83–99.

Caldwell, Mark L. (1979), "*Hamlet* and the senses", *Modern Language Quarterly*, 40: 135–54.

Campbell, Mary Baine (1999), *Wonder and Science: Imagining worlds in early modern Europe* (Ithaca: Cornell University Press).

Cavell, Stanley (2003), *Disowning Knowledge in Seven Plays of Shakespeare* (Cambridge: Cambridge University Press).

Chambers, Douglas (1997), "'Elysium Britannicum not printed neere ready &c': The 'Elysium Britannicum' in the correspondence of John Evelyn" in Therese O'Malley and Joachim Wolschke-Bulmann (eds.), *John Evelyn's "Elysium Britannicum" and European Gardening* (Washington, D.C.: Dumbarton Oaks Research Library and Collection), 107–30.

—— (1992), "'Wild pastorall encounter': John Evelyn, John Beale and the renegotiation of pastoral in the mid-seventeenth century" in Michael Leslie and Timothy Raylor (eds.), *Culture and Cultivation in Early Modern England: Writing and the land* (Leicester and London: Leicester University Press), 173–94.

Christianson, John Robert (2000), *On Tycho's Island: Tycho Brahe and his assistants, 1570–1601* (Cambridge: Cambridge University Press).

Coffin, Charles M. (1937), *John Donne and the New Philosophy* (New York: Columbia University Press, rpt. New York: The Humanities Press, 1958).

Colclough, David (ed.) (2003), *John Donne's Professional Lives* (Cambridge: D. S. Brewer).

Cormack, Lesley B. (1997), *Charting an Empire: Geography at the English universities, 1580–1620* (Chicago: University of Chicago Press).

Cornelius, David K. and Edwin St. Vincent (eds.) (1964), *Cultures in Conflict: Perspectives on the Snow–Leavis controversy* (Chicago: Scott, Foresman and Company).

Corrigan, Brian Jay (2004), *Playhouse Law in Shakespeare's World* (Madison: Farleigh Dickinson University Press).

Daston, Lorraine (ed.) (2000), *Biographies of Scientific Objects* (Chicago: University of Chicago Press).

—— and Katherine Park (1998), *Wonder and the Order of Nature, 1150–1750* (New York: Zone Books).

Dawson, Anthony (2001), "Performance and participation" in Anthony B. Dawson and Paul Yachnin (eds.), *The Culture of Playgoing in Shakespeare's England: A collaborative debate* (Cambridge: Cambridge University Press), 11–37.

Dear, Peter (1995), *Discipline and Experience: The mathematical way in the scientific revolution* (Chicago: University of Chicago Press).

—— (2001), *Revolutionizing the Sciences: European knowledge and its ambitions, 1500–1700* (Princeton: Princeton University Press).

Derrida, Jacques (1994), *Specters of Marx: The state of debt, the work of mourning and the New International*, trans. Peggy Kamuf (New York: Routledge).

Dizikes, Peter (2009), "Our two cultures", *New York Times Book Review*, 22 March, 23.

Drakakis, John (ed.) (1985), *Alternative Shakespeares* (London and New York: Methuen).

Drake, Stillman (1978), *Galileo at Work: His scientific biography* (Chicago: University of Chicago Press).

Eddington, Sir Arthur Stanley (1928), *The Nature of the Physical World* (Cambridge: Cambridge University Press; rpt. Ann Arbor: University of Michigan Press, 1958).

Edgerton, Samuel Y. (1984), "Galileo, Florentine 'Disegno', and the 'strange Spottednesse' of the Moon", *Art Journal*, 44/3: 225–32.

Eliot, T. S. (1950), "Hamlet and his problems" in *Selected Essays* (New York: Harcourt, Brace and Company), 121–6.

Elliott, John R., Jr. (1992), "Drama at the Oxford colleges and the Inns of Court, 1520–1534", *Research Opportunities in Renaissance Drama*, 31: 64–6.

Empson, William (1993), *Essays on Renaissance Literature*, vol. 1: *Donne and the New Philosophy*, ed. John Heffenden (Cambridge: Cambridge University Press).

—— (1951), *The Structure of Complex Words* (London: Chatto and Windus).

Feyerabend, Paul (1975), *Against Method: Outline of an anarchistic theory of knowledge* (London: New Left Books).

Findlen, Paula (2006), "Anatomy theaters, botanical gardens, and natural history collections" in Katharine Park and Lorraine Daston (eds.), *The Cambridge History of Science*, vol. 3: *Early Modern Science* (Cambridge: Cambridge University Press), 272–89.

—— (1994), *Possessing Nature: Museums, collecting, and scientific culture in early modern Italy* (Berkeley: University of California Press).

Fletcher, Angus (2005), "Living magnets, Paracelsian corpses, and the psychology of Grace in Donne's religious verse", *ELH*, 72: 1–22.

—— (2007), *Time, Space, and Motion in the Age of Shakespeare* (Cambridge: Harvard University Press).

Fowler, Alistair (1992), "Georgic and pastoral: Laws of genre in the seventeenth century" in Michael Leslie and Timothy Raylor (eds.), *Culture and Cultivation in Early Modern England: Writing and the land* (Leicester and London: Leicester University Press), 81–8.

Frasca-Spada, Maria and Nichoas Jardine (eds.) (2000), *Books and the Sciences in History* (Cambridge: Cambridge University Press).

Freedberg, David (2002), *The Eye of the Lynx: Galileo, his friends, and the beginnings of modern natural history* (Chicago: University of Chicago Press).

Frost, Kate G. (1990), *Holy Delight: Typology, numerology, and autobiography in Donne's Devotions upon Emergent Occasions* (Princeton: Princeton University Press).

Fuller, Mary C. (1995), *Voyages in Print: English travels to America, 1576–1624* (Cambridge: Cambridge University Press).

Garber, Marjorie (1987), *Shakespeare's Ghost Writers: Literature as uncanny causality* (London: Methuen).

Geertz, Clifford (1966), "The impact of the concept of culture on the concept of man", *Bulletin of the Atomic Scientists*, 22/4: 2–8.

—— (1965), "The impact of the concept of culture on the concept of man" in John R. Platt (ed.), *New Views of the Nature of Man* (Chicago: University of Chicago Press), 93–118.

—— (1980), *Negara: The theater state in nineteenth-century Bali* (Princeton: Princeton University Press).

Gennep, Arnold van (1960), *Rites of Passage*, trans. Monika B. Vizedom and Gabrielle L. Caffe (Chicago: University of Chicago Press).

Gingrich, Own (1981), "Astronomical scrapbook: Great conjunctions, Tycho, and Shakespeare", *Sky and Telescope*, 61: 394–5.

—— and Albert van Helden (2003), "From *Occhiale* to printed page: The making of Galileo's *Sidereus Nuncius*", *Journal for the History of Astronomy*, 34/3: 251–67.

Goldberg, Jonathan (2003), *Shakespeare's Hand* (Minneapolis: University of Minnesota Press).

Gosse, Edmund (1899), *The Life and Letters of John Donne, Dean of St. Paul's* (London; rpt. Gloucester, MA: Peter Smith, 1959).

Green, A. Wigfall (1965), *A Bibliography of the Inns of Court and Chancery* (London: Selden Society).

—— (1931), *The Inns of Court and Early English Drama* (New Haven: Yale University Press, rpt. 1965).

Greenblatt, Stephen (2001), *Hamlet in Purgatory* (Princeton: Princeton University Press).

—— (1991), *Marvelous Possessions: The wonder of the New World* (Chicago: University of Chicago Press).

Haffenden, John (2005), *William Empson: Among the Mandarins* (Oxford: Oxford University Press).

Haraway, Donna (1991), "A Cyborg manifesto: Science, technology, and socialist-feminism in the late twentieth century" in *Simians, Cyborgs, and Women: The reinvention of nature* (New York: Routledge), 149–81.

—— (1988), "Situated knowledges: The science question in feminism and the privilege of partial perspective", *Feminist Studies*, 14/3: 575–600.

——(1991), *Simians, Cyborgs, and Women: The reinvention of nature* (New York: Routledge).

Harkness, Doborah (2007), *The Jewel House: Elizabethan London and the scientific revolution* (New Haven and London: Yale University Press).

Harris, Jonathan Gill (2009), *Untimely Matter in the Time of Shakespeare* (Philadelphia: University of Pennsylvania Press).

Hawkes, Terence (1985), "Telmah" in Patricia Parker and Geoffrey Hartman (eds.), *Shakespeare and the Question of Theory* (New York: Methuen), 312–32.

Hayles, Katherine (1999), *How We Became Posthuman: Virtual bodies in cybernetics, literature and informatics* (Chicago: University of Chicago Press).

Hegel, G. W. F. (1956), *The Philosophy of History*, trans. J. Sibree (New York: Dover Publications).

Heilbron, J. L. (1999), *The Sun in the Church: Churches and solar observatories* (Cambridge: Harvard University Press).

Helgerson, Richard (1992), *Forms of Nationhood: The Elizabethan writing of England* (Chicago: University of Chicago Press).

Henry, John (1997), *The Scientific Revolution and the Origins of Modern Science* (London and New York: Macmillan Press/ St. Martin's Press).

Hotson, Leslie (1938), *I, William Shakespeare* (New York: Oxford University Press).

Howes, David (ed.) (2005), *Empire of the Senses: The sensual culture reader* (Oxford: Berg).

Hunt, John Dixon (1997), "Evelyn's idea of the garden: A theory for all season" in Therese O'Malley and Joachim Wolschke-Bulmahn (eds.), *John Evelyn's "Elysium Britannicum" and European Gardening* (Washington, D.C.: Dumbarton Oaks Research Library and Collection), 269–88.

—— (1976), *The Figure in the Landscape: Poetry, painting, and gardening during the eighteenth century* (Baltimore: Johns Hopkins University Press).

—— (1996), *Garden and Grove: The Italian Renaissance garden in the English imagination, 1600–1750* (Philadelphia: University of Pennsylvania Press).

—— (1992), *Garden and the Picturesque: Studies in the history of landscape architecture* (Cambridge: MIT Press).

—— (ed.) (1992), *Garden History: Issues, approaches, methods* (Washington, D.C.: Dumbarton Oaks Research Library and Collection).

Hunt, John Dixon (2000), *Greater Perfections: The practice of garden theory* (Philadelphia: University of Pennsylvania Press).

Impey, Oliver and Arthur MacGregor (eds.) (1985), *The Origins of Museums: The cabinet of curiosities in sixteenth- and seventeenth-century Europe* (Oxford: Oxford University Press).

Jardine, Lisa (1974), *Francis Bacon: Discovery and the art of discourse* (Cambridge: Cambridge University Press).

Johns, Adrian (2000), *The Nature of the Book: Print and knowledge in the making* (Chicago: University of Chicago Press).

Johnson, David E. (2001), "Descartes's *Corps*", *Arizona Quarterly*, 57/1: 113–52.

Keller, Evelyn Fox (1985), *Reflections on Gender and Science* (New Haven: Yale University Press).

Keynes, Geoffrey (1968), *John Evelyn: A study in bibliophily* (Oxford: Clarendon Press).

Knapp, Margaret and Michal Kobialka (1984), "Shakespeare and the Prince of Purpoole: The 1594 production of *The Comedy of Errors* at Gray's Inn Hall", *Theatre History Studies*, 4: 71–81.

Kuchar, Gary (2005), *Divine Subjection: The rhetoric of sacramental devotion in early modern England* (Pittsburgh: Duquesne University Press).

Kuhn, Thomas S. (1996), *The Structure of Scientific Revolutions*, 3rd edn (Chicago: University of Chicago Press).

Labinger, Jay A. and Harry Collins (eds.) (2001), *The One Culture? A conversation about science* (Chicago and London: The University of Chicago Press).

Lacan, Jacques (1978), *The Four Fundamental Concepts of Pyscho-analysis*, ed. Jacques-Alain Miller and tr. Alan Sheridan (New York and London: W. W. Norton and Company).

Lanier, Douglas (1993), "'Stigmatical in making': The material character of *The Comedy of Errors*", *English Literary Renaissance*, 23/1: 81–112.

Latour, Bruno (1992), "One more turn after the social turn . . . " in Ernan McMullan (ed.), *The Social Dimension of Science* (South Bend: University of Notre Dame Press), 272–94.

—— (1999), *Pandora's Hope: Essays on the reality of science studies* (Cambridge: Harvard University Press).

—— (2004), *Politics of Nature: How to bring the sciences into democracy*, trans. C. Porter (Cambridge: Harvard University Press).

—— (1987) *Science in Action: How to follow scientists and engineers through society* (Cambridge: Harvard University Press).

—— (1993) *We Have Never Been Modern*, trans. C. Porter (Cambridge: Harvard University Press).

—— and Steve Woolgar (1986), *Laboratory Life: The construction of scientific facts* (Princeton: Princeton University Press).

Lattis, James M. (1994), *Between Copernicus and Galileo: Christoph Clavius and the collapse of the Ptolemaic cosmology* (Chicago: University of Chicago Press).

Leavis, F. R. (1962), *Two Cultures?: The significance of C. P. Snow* (London: Chatto and Windus).

Leith, Sam (2009), "Science friction", *Financial Times Weekend*: "Life and Arts", 10 May, 1.

Lenoir, Timothy (1994), "Was the last turn the right turn?: The semiotic turn and A. J. Greimas", *Configurations*, 1: 119–36.

Leslie, Michael and Timothy Raylor (eds.) (1992), *Culture and Cultivation in Early Modern England: Writing and the land* (Leicester and London: Leicester University Press).

Levine, George (ed.) (1987), *One Culture: Essays in science and literature* (Madison: University of Wisconsin Press).

Lindberg, David C. (1990), "Conceptions of the scientific revolution from Bacon to Butterfield: A preliminary sketch" in David C. Lindberg and Robert S. Westman (eds.), *Reappraisals of the Scientific Revolution* (Cambridge: Cambridge University Press), 1–26.

—— and Robert S. Westman (eds.) (1990), *Reappraisals of the Scientific Revolution* (Cambridge: Cambridge University Press).

Low, Anthony (1985), *The Georgic Revolution* (Princeton: Princeton University Press).

Lumsden, Charles and E. O. Wilson (1981), *Genes, Mind, and Culture: The coevolutionary process* (Cambridge: Harvard University Press).

—— (1983), *Promethean Fire: Reflections on the origin of mind* (Cambridge: Harvard University Press).

Magnusson, Lynne A. (2004), "Scoff power in *Love's Labour's Lost* and the Inns of Court: Language in context", *Shakespeare Survey*, 57: 196–208.

Maisano, Scott (2007), "Infinite gesture: Automata and the emotions in Descartes and Shakespeare" in Jessica Riskin (ed.), *Genesis Redux: Essays in the history and philosophy of artificial life* (Chicago and London: University of Chicago Press), 63–84.

McMullan, E. (ed.) (1992), *The Social Dimension of Science* (Notre Dame: University of Notre Dame Press).

McRae, Andrew (1996), *God Speed the Plough: The representation of agrarian England, 1500–1660* (Cambridge: Cambridge University Press).

Marchitello, Howard (2003), "Garden *frisson*", *Journal of Medieval and Early Modern Studies*, 33/1: 143–77.

—— (1997), *Narrative and Meaning in Early Modern England: Browne's skull and other histories* (Cambridge: Cambridge University Press).

Marx, Leo (1964), *The Machine in the Garden: Technology and the pastoral idea in America* (Oxford: Oxford University Press; rpt. 2000).

Maus, Katherine Eisaman (1985), *Inwardness and Theater in the English Renaissance* (Chicago: University of Chicago Press).

Mazzio, Carla (2009), "Shakespeare and science", *South Central Review*, 2/1&2: 1–23.

—— (2005), "The senses divided: Organs, objects, and media in early modern England" in David Howes (ed.) *Empire of the Senses: The sensual culture reader* (Oxford: Berg), 85–105.

Miller, Mara (1993), *The Garden as an Art* (Albany: State University of New York Press).

Mueller, Janel (1968), "The exegesis of experience: Dean Donne's *Devotions upon Emergent Occasions*", *Journal of English and Germanic Philology*, 67: 1–19.

Nicolson, Marjorie (1950), *The Breaking of the Circle: Studies in the effect of the "new science" upon seventeenth-century poetry* (Evanston: Northwestern University Press).

—— (1946), *Newton Demands the Muse: Newton's Opticks and the eighteenth-century poets* (Princeton: Princeton University Press).

—— (1965), *Pepys' Diary and the New Science* (Charlottesville: University of Virginia Press).

—— (1956), *Science and Imagination* (Ithaca: Great Seal Books/ Cornell University Press, rpt. 1962).

O'Callaghan, Michelle (2007), *The English Wits: Literature and sociability in early modern England* (Cambridge: Cambridge University Press).

Orgel, Stephen (1975), *The Illusion of Power: Political theater in the English Renaissance* (Berkeley: University of California Press).

—— (1965), *The Jonsonian Masque* (Cambridge: Harvard University Press).

Park, Katharine and Lorraine Daston (2006), "Introduction: The age of the new" in *Early Modern Science, The Cambridge History of Science*, vol. 3 (Cambridge: Cambridge University Press), 1–17.

Paster, Gail Kern (1993), *The Body Embarrassed: Drama and the Disciplines of shame in early modern England* (Ithaca: Cornell University Press).

—— (2004), *Humoring the Body: Emotions and the Shakespearean stage* (Chicago: University of Chicago Press).

Paster, Gail Kern, Katherine Rowe, and Mary Floyd-Wilson (eds.) (2004), *Reading the Early Modern Passions: Essays in the cultural history of emotion* (Philadelphia: University of Pennsylvania Press).

Pender, Stephen (2003), "Essaying the body: Donne, affliction, and medicine" in David Colclough (ed.), *John Donne's Professional Lives* (Cambridge: D. S. Brewer), 215–48.

Peterfreund, Stuart (2008), "From the Forbidden to the Familiar: The way of Natural Theology Leading up to and beyond the Long Eighteenth Century", *Studies in Eighteenth-Century Culture*, 37: 23–39.

—— (2000), "Imagination at a Distance: Bacon's Epistemological Double-Bind, Natural Theology, and the Way of Scientific Explanation in the Seventeenth and Eighteenth Centuries", *Eighteenth Century: Theory and Interpretation*, 41/2: 110–40.

Pickering, Andrew (1995), *The Mangle of Practice: Time, Agency, and Science* (Chicago: University Press).

Platt, John R. (ed.) (1965), *New Views of the Nature of Man* (Chicago: University of Chicago Press).

Porter, Roy (1996), "The two cultures revisited", *boundary 2*, 23/2: 1–17.

Prest, John (1981), *The Garden of Eden: The botanic garden and the re-creation paradise* (New Haven: Yale University Press).

Prest, Wilfred R. (1972), *The Inns of Court Under Elizabeth I and the Early Stuarts, 1590–1640* (Totowa, NJ: Rowman and Littlefield).

Preston, Claire (2000), "In the wilderness of forms: Ideas and things in Thomas Browne's cabinets of curiosity" in Neil Rhodes and Jonathan Sawday (eds.), *The Renaissance Computer: Knowledge technology in the first age of print* (London: Routledge), 170–83.

—— (2006), "Of cyder and sallets: The hortulan saints and *The Garden of Cyrus*", *Literature Compass*, 3/4: 867–83.

—— (2005), *Thomas Browne and the Writing of Early Modern Science* (Cambridge: Cambridge University Press).

Pumfrey, Stephen, Paolo L. Rossi, and Maurice Slawinski (eds.) (1991), *Science, Culture and Popular Belief in Renaissance Europe* (Manchester: Manchester University Press).

Raffield, Paul (2004), *Images and Cultures of Law in Early Modern England: Justice and political power, 1558–1660* (Cambridge: Cambridge University Press).

Reeves, Eileen (1997), *Painting the Heavens: Art and science in the age of Galileo* (Princeton: Princeton University Press).

Renn, Jürgen (ed.) (2001), *Galileo in Context* (Cambridge: Cambridge University Press).

Rheinberger, Hans-Jörge (1997), *Toward a History of Epistemic Things: Synthesizing proteins in the test tube* (Stanford: Stanford University Press).

Rhodes, Neil and Jonathan Sawday (2000), *The Renaissance Computer: Knowledge technology in the first age of print* (London: Routledge).

Riskin, Jessica (ed.) (2007), *Gensis Redux: Essays in the history and philosophy of artificial life* (Chicago and London: University of Chicago Press).

Rivlin, Elizabeth (2002), "Theatrical literacy in *The Comedy of Errors* and the *Gesta Grayorum*", *Critical Survey*, 14/1: 64–78.

Roberts, John R (1973–82), *John Donne: An annotated bibliography of modern criticism*, vol. 1: *1912–1978* (Columbia: University of Missouri Press) and (2004), vol. 2: *1979–1995* (Pittsburgh: Duquesne University Press).

Rose, Jacqueline (1985), "Sexuality in the reading of Shakespeare: *Hamlet* and *Measure for Measure*" in John Drakakis (ed.), *Alternative Shakespeares* (London and New York: Methuen), 95–118.

Ross, Stephanie (1998), *What Gardens Mean* (Chicago: University of Chicago Press).

Rossi, Paolo (1978), *Francis Bacon: From magic to science*, trans. S. Rabinovitch (London: Routledge and Kegan Paul).

Ryle, Gilbert (1949), *The Concept of Mind* (London: Hutchinson and Co.).

Sawday, Jonathan (2007), *Engines of the Imagination: Renaissance culture and the rise of the machine* (London and New York: Routledge).

Scarry, Elaine (1988), "Donne: 'But yet the body is his booke'" in *Literature and the Body: Essays on populations and persons* (Baltimore: Johns Hopkins University Press), 70–105.

—— (ed.) (1988), *Literature and the Body: Essays on populations and persons* (Baltimore: Johns Hopkins University Press).

Schoenfeldt, Michael (1999), *Bodies and Selves in Early Modern England: Physiology and inwardness in Spenser, Shakespeare, Herbert, and Milton* (Cambridge: Cambridge University Press).

Serres, Michel (1989), *Detachment*, trans. G. James and R. Federman (Athens: Ohio University Press).

—— (1982), *Hermes: Literature, science, philosophy*, trans. J. V. Harari and D. F. Bell (Baltimore: Johns Hopkins University Press).

—— (ed.) (1995a), *A History of Scientific Thought: Elements of a History of Science* (Oxford: Blackwell).

—— (1995b), *The Natural Contract*, trans. E. MacArthur and W. Paulson (Ann Arbor: University of Michigan Press).

—— (1997), *The Troubadour of Knowledge*, trans. S. F. Glaser and W. Paulson (Ann Arbor: University of Michigan Press).

—— and Bruno Latour (1995), *Conversations on Science, culture, and time*, trans. Roxanne Lapidus (Ann Arbor: University of Michigan Press).

Shapin, Steven (1994), *A Social History of Truth: Civility and science in seventeenth-century England* (Chicago: University of Chicago Press).

—— and Simon Schaffer (1985), *Leviathan and the Air-Pump: Hobbes, Boyle, and the experimental life* (Princeton: Princeton University Press).

Shapiro, Barbara (2000), *A Culture of Fact: England, 1550–1720* (Ithaca: Cornell University Press).

Singer, Charles (1941), *A Short History of Science to the Nineteenth Century* (Oxford: The Clarendon Press).

Smith, Bruce R. (2004), "Hearing green" in G. K. Paster, K. Rowe, and M. Floyd-Wilson (eds.), *Reading the Early Modern Passions: Essays in the cultural history of emotion* (Philadelphia: University of Pennsylvania Press), 147–68.

—— (1997), "A night of errors and the dawn of empire: Male enterprise in *The Comedy of Errors*" in Michael J. Collins (ed.), *Shakespeare's Sweet Thunder: Essays on the early comedies* (Newark: University of Delaware Press), 102–25.

Snow, C. P. (1988), *The Two Cultures*, ed. Stefan Collini (Cambridge: Cambridge University Press).

—— (1959), *The Two Cultures and the Scientific Revolution* (Cambridge: Cambridge University Press).

Sommerville, Johann P. (2003), "John Donne the controversialist: The poet as political thinker" in David Colclough (ed.), *John Donne's Professional Lives* (Cambridge: D. S. Brewer), 73–95.

Spiller, Elizabeth (2004), *Science, Reading, and Renaissance Literature: The Art of making knowledge, 1580–1670* (Cambridge: Cambridge University Press).

—— (2009), "Shakespeare and the making of early modern science: Resituating Prospero's art", *South Central Review*, 26/1&2: 24–41.

Stallybrass, Peter (2001), "Hauntings: The materiality of memory on the Renaissance stage" in Valeria Finucci and Kevin Brownlee (eds.), *Generation and Degeneration: Tropes of reproduction in literature and history from antiquity through early modern Europe* (Durham: Duke University Press), 287–316.

—— and Roger Chartier, J. Frankline Mowrey, and Heather Wolfe (2004), "Hamlet's tables and the technologies of writing in Renaissance England", *Shakespeare Quarterly*, 55: 379–419.

Stewart, Susan (1998), "Garden agon", *Representations*, 62: 111–43.

Sugg, Richard (2007), *John Donne* (Houndsmills and New York: Palgrave Macmillan).

Swann, Marjorie (2001), *Curiosity and Texts: The culture of collecting in early modern England* (Philadelphia: University of Pennsylvania Press).

Targoff, Ramie (2006), "Facing death" in Achsah Guibbory (ed.), *The Cambridge Companion to John Donne* (Cambridge: Cambridge University Press), 217–31.

—— (2008), *John Donne, Body and Soul* (Chicago: University of Chicago Press).

Thoren, Victor E., with John R. Christianson (1990), *The Lord of Uraniborg: A biography of Tycho Brahe* (Cambridge: Cambridge University Press).

Tigner, Amy L. (2006), "*The Winter's Tale*: Gardens and the marvels of transformation", *English Literary Renaissance*, 36: 114–34.

Traub, Valerie (2002), *The Renaissance of Lesbianism in Early Modern England* (Cambridge: Cambridge University Press).

Turner, Henry S. (2006), *The English Renaissance Stage: Geometry, poetics, and the practical spatial arts, 1580–1630* (Oxford: Oxford University Press).

—— (2009), 'Life science: Rude mechanicals, human mortals, posthuman Shakespeare', *South Central Review* 26/1&2: 197–217.

—— (2007), *Shakespeare's Double Helix* (London: Continuum).

Turner, Victor (1982), *From Ritual to Theatre: The human seriousness of play* (New York: Performing Arts Journal Publications).

Usher, Peter (2001), "Advances in the Hamlet cosmic allegory", *The Oxfordian*, 5: 25–49.

—— (2005), "Hamlet's love letter and the new astronomy", *The Oxfordian*, 8: 93–109.

—— (1999), "Hamlet's transformation", *Elizabethan Review*, 7: 48–64.

Vickers, Brian (ed.) (1984), *Occult and Scientific Mentalities in the Renaissance* (Cambridge: Cambridge University Press).

Watson, Robert N. (2006), *Back to Nature: The green and the real in the late Renaissance* (Philadelphia: University of Pennsylvania Press).

Waugh, Patricia (1999), "Revising the two cultures debate: Science, literature, and value" in David Fuller and Patricia Waugh (eds.), *The Arts and Sciences of Criticism* (Oxford: Oxford University Press), 33–59.

West, William N. (2008), "'But this will be a mere confusion': Real and represented confusions on the Elizabethan stage", *Theatre Journal* 60/2: 217–33.

Wilson, E. O. (1998), *Consilience: The unity of knowledge* (New York: Knopf).

Wimsatt, W. K. (1954), *The Verbal Icon: Studies in the meaning of poetry* (Lexington: University Press of Kentucky).

—— and Monroe C. Beardsley (1946), "The intentional fallacy", *Sewanee Review*, 54: 468–88.

Winkler, Mary G. and Albert van Helden (1992), "Representing the heavens: Galileo and visual astronomy", *Isis*, 83/2: 195–217.

Wolfe, Jessica (2004), *Humanism, Machinery, and Renaissance Literature* (Cambridge: Cambridge University Press).

Wolin, Richard (1982), *Walter Benjamin: An aesthetics of redemption* (Berkeley: University of California Press, rpt. 1994).

Index